Holt
Mathematics

Chapter 1 Resource Book

HOLT, RINEHART AND WINSTON

A Harcourt Education Company

Orlando • **Austin** • New York • San Diego • London

ISBN 0-03-078191-4

7 8 9 10   170   10 09

CONTENTS

Holt Mathematics

Holt Mathematics

Date _____

Dear Family,

In this chapter, your child will learn about comparing and ordering whole numbers, about estimating with whole numbers, and about exponents.

Comparing and ordering whole numbers helps your child read and understand large numbers as they are typically used in science, business, and technology. We can learn to compare and order whole numbers by using place value or a number line. We designate the number order from least to greatest by using symbols. The symbol < means "less than," and > means "greater than," so that 1,266 > 835 and 835 < 1,266.

To order numbers, we can compare them using place value, then write them in order from least to greatest. Here is an example of using place value to compare whole numbers:

Belgium's population in 2005 was 10,364,388 people. The Czech Republic's population in 2005 was 10,241,138 people. Which country had more people in 2005?

1 0 , 3 6 4 , 3 8 8   *Start at the left and compare digits in the same place value position.*

1 0 , 2 4 1 , 1 3 8   *Look for the first place the values are different.*

200 thousand < 300 thousand
10,241,138 < 10,364,388
Belgium had more people.

We can also use a number line to order whole numbers, as in the following example:

Write the numbers in order from least to greatest.
923; 835; 1,266

Graph each number on a number line.

923 is between 900 and 1,000
835 is between 800 and 900
1,266 is between 1,200 and 1,300

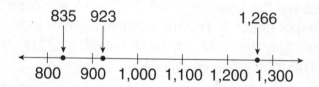

The numbers are ordered when you read the number line from left to right. In order from least to greatest they are 835, 923, and 1,266.

Holt Mathematics

Estimating with whole numbers will help your child do math mentally when an exact answer is not needed. One way to estimate whole numbers is by rounding. Look at the digit to the right of the place that you are rounding. If the digit is 5 or greater, round upward in value. If the digit is less than 5, round down in value. This basic guideline will help you avoid overestimating or underestimating.

For example, you can estimate a sum by rounding to a certain place value.

Estimate 5,439 + 7,516 by rounding to thousands.

$$
\begin{array}{r}
5,000 \\
+\ 8,000 \\
\hline
13,000
\end{array}
$$

The sum is about 13,000.

You can also round the numbers in a problem to **compatible numbers,** which are numbers that are close to the numbers in the problem.

Mrs. Byrd will drive 122 miles to take Becca to the state fair. She can drive 65 mi/h. About how long will the trip take?

To find out how long the trip will be, divide the miles Mrs. Byrd has to travel by how many miles per hour she can drive.

miles ÷ miles per hour

$122 \div 65 \longrightarrow 120 \div 60$ 120 and 60 are compatible because they are easy to divide mentally.

$120 \div 60 = 2$

It will take Mrs. Byrd about 2 hours to reach the state fair.

Using exponents is a more efficient way of expressing large numbers. They typically appear, for example, in scientific publications and communications and in calculations involving large numbers. An exponent tells how many times to multiply a number, called the **base,** by itself. In 7^3 feet, the base 7 is written with an exponent of 3. The exponential number indicates how many times to multiply the base number by itself, so 7^3 is the same as $7 \times 7 \times 7$. The total value equals 343.

For additional resources, visit go.hrw.com and enter the keyword MR7 Parent.

Holt Mathematics

Name _____ Date _____ Class _____

Practice A
Comparing and Ordering Whole Numbers

Write <, >, or = to compare the numbers.

1. 8 ☐ 18   **2.** 43 ☐ 34   **3.** 100 ☐ 90

4. 295 ☐ 259   **5.** 706 ☐ 706   **6.** 1,006 ☐ 6,001

Write the numbers from least to greatest.

7. 3; 13; 1   **8.** 88; 80; 78   **9.** 104; 204; 102

_____   _____   _____

10. 75; 95; 59   **11.** 642; 855; 658   **12.** 274; 207; 740

_____   _____   _____

Write the numbers from greatest to least.

13. 10; 100; 11   **14.** 36; 16; 63   **15.** 28; 20; 80

_____   _____   _____

16. 500; 300; 305   **17.** 593; 93; 59   **18.** 184; 800; 481

_____   _____   _____

19. English is spoken in 47 countries around the world. French is spoken in 23 countries. Which language is spoken in the most countries?

_____

20. The United States–Mexico border is 1,933 miles long. The United States–Canada border is 3,987 miles long. Which border is longer?

_____

Holt Mathematics

LESSON 1-1 Practice B
Comparing and Ordering Whole Numbers

Compare. Write <, >, or =.

1. 69 ☐ 96

2. 117 ☐ 107

3. 958 ☐ 9,124

4. 3,567 ☐ 3,567

5. 18,443 ☐ 1,844

6. 64,209 ☐ 64,290

Order the numbers from least to greatest.

7. 58; 166; 85

8. 115; 151; 111

9. 269; 29; 96

_____

_____

_____

10. 308; 3,800; 3,080

11. 1,864; 824; 1,648

12. 4,663; 4,336; 43,666

_____

_____

_____

Order the numbers from greatest to least.

13. 35; 53; 13

14. 807; 800; 708

15. 249; 392; 248

_____

_____

_____

16. 555; 600; 535

17. 7,320; 6,000; 6,305

18. 999; 9,559; 5,995

_____

_____

_____

19. Delaware and Rhode Island are the two smallest states. Delaware covers 1,955 square miles, and Rhode Island covers 1,045 square miles. What is the smallest state in the United States?

_____

20. Vermont and Wyoming have the smallest populations in the United States. The population of Vermont is 608,827. The population of Wyoming is 493,782. Which state has the smallest population?

_____

Holt Mathematics

LESSON 1-1 Practice C
Comparing and Ordering Whole Numbers

Compare. Write <, >, or =.

1. 1,478 ☐ 1,748

2. 5,643 ☐ 5,643

3. 9,610 ☐ 10,961

4. 308,524 ☐ 3,854

Order the numbers from least to greatest.

5. 379; 79; 978

6. 16,780; 17,847; 6,988

_____

_____

7. 76,334; 47,961; 70,336

8. 101,695; 19,568; 191,658

_____

_____

Order the numbers from greatest to least.

9. 605; 560; 565

10. 8,320; 8,063; 8,663

_____

_____

11. 49,210; 49,000; 49,910

12. 352,699; 353,963; 95,614

_____

_____

13. Alaska, California, and Texas are the three largest states. Alaska covers 615,230 square miles. California covers 158,869 square miles. Texas covers 267,277 square miles. Write the states in order by size, from largest to smallest.

_____

14. California, New York, and Texas have the largest populations in the United States. Their populations are 33,871,648; 20,851,820; and 18,976,457. California has the largest population. More people live in Texas than in New York. What is each state's population?

_____

Holt Mathematics

LESSON 1-1 Reteach
Comparing and Ordering Whole Numbers

You can use place value to compare or order whole numbers.
Use < or > to compare the numbers.

289,865 ☐ 289,765

Thousands			Ones			Compare the digits
H	**T**	**O**	**H**	**T**	**O**	from left to right.
2	8	9	**8**	6	5	First Number
2	8	9	**7**	6	5	Second Number

8 > 7
So, 289,865 > 289,765

Write < or > to compare the numbers.

1.

Thousands			Ones		
H	**T**	**O**	**H**	**T**	**O**
		3	5	4	7
		3	5	3	2

3,547 ☐ 3,532

2.

Thousands			Ones		
H	**T**	**O**	**H**	**T**	**O**
		9	5	3	6
		9	6	3	5

9,536 ☐ 9,635

Write the numbers in order from least to greatest.
976; 859; 924

Ones		
H	**T**	**O**
9	7	6
8	5	9
9	2	4

Compare the numbers in pairs.
976 > 859, 976 > 924, and 859 < 924.
So the numbers from least to greatest are
859; 924; 976.

Write the numbers in order from least to greatest.

3.

Ones		
9	5	4
9	4	5
9	6	9

4.

Ones		
3	4	3
3	3	4
4	3	4

5.

Ones		
8	9	4
8	9	2
9	6	5

Holt Mathematics

LESSON 1-1 **Challenge**
Ancient Calculators

Long before place-value charts were invented, people used a tool
called an abacus to show and read numbers. The Chinese began
using the *Suan Pan* abacus, shown below, about 800 years ago.

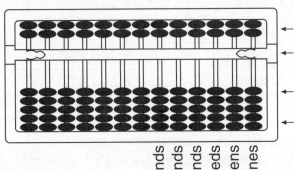

← Each bead above the center
bar stands for 5 units.

← Center bar

← Each bead below the center
bar stands for 1 unit.

← Each rod, or row, of beads
represents one place value.

hundred thousands
ten thousands
thousands
hundreds
tens
ones

To use an abacus to show and read a number, move the beads
to the center bar for each place value and add. For example:

2 8 4 0 5 = 28,405

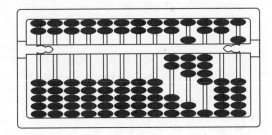

6 3 1 7 0 9 = 631,709

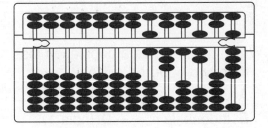

Write the number shown on each abacus.
Then write < or > to compare the numbers.

1.

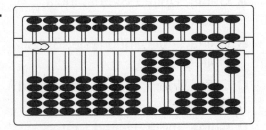

2.

_____   _____

Holt Mathematics

Problem Solving

LESSON 1-1

Comparing and Ordering Whole Numbers

Use the tables below to answer each question.

Most Populated Countries	
Brazil	174,468,575
China	1,273,111,290
India	1,029,991,145
Indonesia	228,437,870
United States	278,058,881

Largest Countries (square mi)	
Brazil	3,265,059
Canada	3,849,646
China	3,705,408
Russia	6,592,812
United States	3,539,224

1. Which country has the greatest population?

   _____

2. Which countries have more than one billion people?

   _____

3. Which country is the largest in the world?

   _____

4. Which country's area is closest to 4,000,000 square miles?

   _____

5. What is the error in the following statement? Canada is larger than the United States, but smaller than China.

   _____

   _____

6. Based on population and size, which country do you think is more crowded, Brazil or the United States? Explain.

   _____

   _____

7. Which country has a population less than two hundred million?

   A China   **C** Brazil

   B Indonesia   **D** India

8. Which countries have populations greater than the United States?

   F China and Brazil

   G China and India

   H India and Indonesia

   J Indonesia and China

9. Which list shows the countries in order by population from greatest to least?

   A China, United States, India, Indonesia, Brazil

   B China, India, Indonesia, Brazil, United States,

   C China, India, Indonesia, United States, Brazil

   D China, India, United States, Indonesia, Brazil

10. Which list shows the countries in order by size from smallest to largest?

   F Brazil, United States, China, Canada, Russia

   G Brazil, United States, Canada, China, Russia

   H Brazil, United States, Canada, Russia, China

   J Brazil, United States, Russia, China, Canada

Holt Mathematics

LESSON 1-1 Reading Strategies
Analyze Information

Reading numbers helps you compare them.

Number	Read
2,581	2 thousand, 581
6,328	6 thousand, 328

Compare greatest place value: thousands.
6 thousand is greater than 2 thousand. So,

6,328 > 2,581 or 2,581 < 6,328.

Larger numbers can be compared and ordered in the same way.

Number	Read
453,276,328	453 million, 276 thousand, 328
435,617,119	435 million, 617 thousand, 119
457,428,937	457 million, 428 thousand, 937

Compare greatest place value: millions.
457 million > 453 million > 435 million

These three numbers in order from greatest to least are
457,428,937; 453,276,328; 435,617,119.

Use the numbers 637,598 and 673,522 to answer Exercises 1–3.

1. Write how these two numbers are read.

   637,598 _____

   673,522 _____

2. Which place value will you compare to decide
   which number is greater? _____

3. Use > or < to compare. 637,598 ☐ 673,522

Use the numbers 353,276,128; 353,268,437; and 353,248,753 to
answer Exercises 4–6.

4. Write how these numbers are read.

   353,276,128 _____

   353,268,437 _____

   353,248,753 _____

5. Which place value will you compare to put the
   numbers in order? _____

6. Order these three numbers from least to greatest.

   _____

Holt Mathematics

Puzzles, Twisters & Teasers

LESSON 1-1

Did you know?

What is one way tarantulas defend themselves?

Let's find out! Circle the greater number. Put the letter above
the greater number in the boxes below.

	T	L			W	H
1.	691	619		11.	99,987	100,000

	S	H			U	A
2.	5,618	8,567		12.	5,009	5,509

	E	I			W	I
3.	2,649	2,469		13.	10,080	10,800

	S	Y			R	Q
4.	957	975		14.	687	678

	L	C			A	B
5.	13,485	12,635		15.	1,013	1,173

	A	I			E	A
6.	873	854		16.	213,946	214,026

	E	U			L	V
7.	512,009	512,125		17.	1,217	1,127

	T	N			M	L
8.	308	380		18.	917	971

	C	K			S	N
9.	7,498	7,398		19.	2,913	2,513

	H	J				
10.	913,003	912,999				

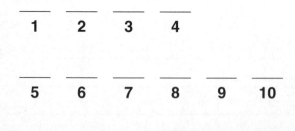

___	___	___	___
1	2	3	4

___	___	___	___	___	___
5	6	7	8	9	10

___	___	___	___	___	___	___	___	___ |.
11	12	13	14	15	16	17	18	19

Holt Mathematics

Practice A

LESSON 1-2

Estimating with Whole Numbers

Round each number to the greatest place value.

1. 67 _____

2. 81 _____

3. 24 _____

4. 115 _____

5. 575 _____

6. 1,852 _____

Estimate each sum or difference.

7. 42 + 19

8. 63 − 28

9. 37 + 34

_____

_____

_____

10. 93 − 14

11. 104 + 178

12. 112 − 9

_____

_____

_____

Estimate each product.

13. 2 × 19

14. 87 × 2

15. 26 × 3

_____

_____

_____

Rewrite each problem using compatible numbers. Then divide.

16. 148 ÷ 5

17. 412 ÷ 4

18. 70 ÷ 6

_____

_____

_____

19. 62 ÷ 3

20. 40 ÷ 7

21. 29 ÷ 4

_____

_____

_____

22. A fin whale weighs 44 tons. A gray whale weighs 32 tons. About how much more does a fin whale weigh than a gray whale?

_____

23. The Suez Canal in Egypt is 108 miles long. The Erie Canal in New York is 363 miles long. About how long are the two canals together?

_____

11

Holt Mathematics

Practice B
Estimating with Whole Numbers

Estimate each sum or difference.

1. 67 + 14

2. 583 − 329

3. 94 − 36

_____

_____

_____

4. 2,856 + 2,207

5. 276 + 316

6. 6,020 − 3,688

_____

_____

_____

7. 34,465 + 19,002

8. 78,135 − 19,431

9. 216,135 + 165,800

_____

_____

_____

Estimate each product or quotient.

10. 59 ÷ 6

11. 51 × 8

12. 83 ÷ 4

_____

_____

_____

13. 9 × 27

14. 49 ÷ 6

15. 53 × 8

_____

_____

_____

16. 147 ÷ 5

17. 118 ÷ 6

18. 79 × 5

_____

_____

_____

19. Sailfish are the fastest fish in the world. They can swim 68 miles an hour. About how far can a sailfish swim in 3 hours?

_____

20. At a height of 3,281 feet, Angel Falls in Venezuela is the tallest waterfall in the world. Niagara Falls in the United States is only 190 feet tall. About how much taller is Angel Falls?

_____

21. Ali, a gardener, is preparing to fertilize a lawn. The lawn is 30 yards by 25 yards. One bag of fertilizer will cover an area of 100 square yards. How many bags of fertilizer does Ali need to buy?

_____

Holt Mathematics

Practice C
Estimating with Whole Numbers

Estimate each sum or difference.

1. 651 + 124

2. 344 − 175

3. 1,862 + 1,403

_____

_____

_____

4. 25,661 + 11,706

5. 59,210 − 24,337

6. 542,901 + 251,504

_____

_____

_____

7. 346,132 − 131,649

8. 292,126 + 167,165

9. 912,910 − 315,904

_____

_____

_____

Estimate each product or quotient.

10. 76 × 3

11. 124 ÷ 3

12. 57 × 4

_____

_____

_____

13. 538 ÷ 61

14. 359 ÷ 64

15. 179 × 21

_____

_____

_____

16. 8 × 56

17. 263 ÷ 13

18. 9 × 63

_____

_____

_____

19. The greatest depth of the Sea of Japan is 12,276 feet. The Bering Sea is 3,383 feet deeper than the Sea of Japan. The Caribbean Sea is 7,129 feet deeper than the Bering Sea. About how deep is the Bering Sea? the Caribbean Sea?

_____

20. Sperm whales dive up to 7,476 feet in search of food, which is about 9 times deeper than emperor penguins dive. About how deep do the penguins dive?

_____

Holt Mathematics

Reteach
LESSON 1-2

Estimating with Whole Numbers

In mathematics, you can find an estimate when an exact answer is
not needed. An estimate is close to the exact answer.

You can use rounding to estimate sums and differences.

A. Estimate the sum by rounding to
the hundreds.

$$3,478 \longrightarrow \quad 3,500$$
$$+\ 7,136 \longrightarrow +\ 7,100$$
$$\overline{\ 10,600}$$

B. Estimate the difference by rounding
to the thousands.

$$23,848 \longrightarrow \quad 24,000$$
$$-\ 16,132 \longrightarrow -\ 16,000$$
$$\overline{\ 8,000}$$

**Estimate each sum or difference by rounding to the place value
indicated.**

1. hundreds

$$789 \longrightarrow$$
$$+\ 453 \longrightarrow +$$

2. thousands

$$4,987 \longrightarrow$$
$$-\ 2,348 \longrightarrow -$$

3. tens

$$456 \longrightarrow$$
$$+\ 875 \longrightarrow +$$

4. tens

$$876 \longrightarrow$$
$$-\ 432 \longrightarrow -$$

5. hundreds

$$6,898 \longrightarrow$$
$$+\ 2,671 \longrightarrow +$$

6. thousands

$$1,857 \longrightarrow$$
$$+\ 3,598 \longrightarrow +$$

7. hundreds

$$8,813 \longrightarrow$$
$$-\ 2,384 \longrightarrow -$$

8. thousands

$$9,128 \longrightarrow$$
$$-\ 4,716 \longrightarrow -$$

Holt Mathematics

Reteach

LESSON 1-2 *Estimating with Whole Numbers (continued)*

You can use rounding and basic facts to estimate products. Count the number of zeros in your rounded numbers. They will appear to the right of the basic fact in your estimate.

Estimate 8 × 532.

8 × 532
↓   ↓
8 × 500 Round each factor.

↓   ↓   two zeros

4,000

Use rounding to estimate each product.

9. 28 × 5

10. 78 × 11

11. 67 × 19

12. 93 × 7

_____   _____   _____   _____

Compatible numbers are numbers that are easy to compute mentally. One compatible number divides evenly into the other.

Estimate the quotient of 553 ÷ 8.

Step 1: What are the multiples of 8?
8 16 24 32 40 48 56 64
Which multiple is closest to 55?
56 is close to 55.
8 and 560 are compatible numbers.

Step 2: Divide. 560 ÷ 8 = 70

Use compatible numbers to estimate each quotient.

13. 748 ÷ 25   **14.** 557 ÷ 8   **15.** 417 ÷ 7   **16.** 241 ÷ 3

_____   _____   _____   _____

Holt Mathematics

Challenge

A Shopping Spree!

You have just won a $2,000 shopping spree at Electronics City!
Use estimation and the store's advertisement below to make
two different shopping lists of what you can buy without going
over your spending limit.

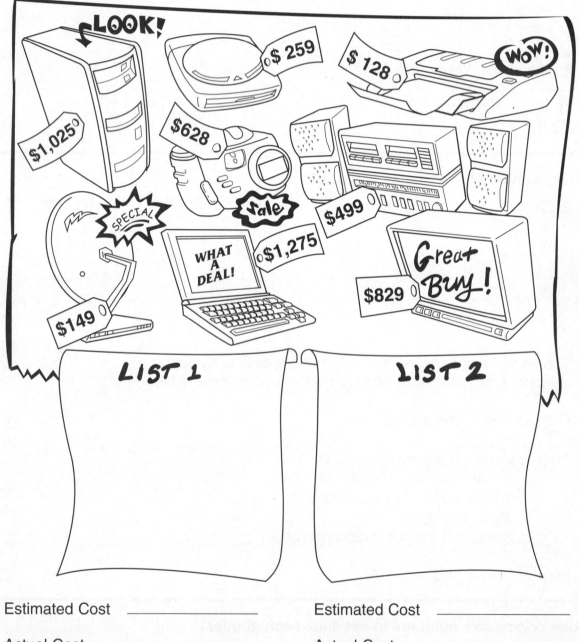

Estimated Cost _____   Estimated Cost _____

Actual Cost _____   Actual Cost _____

Holt Mathematics

Problem Solving

LESSON 1-2

Estimating with Whole Numbers

Use the table below to answer each question.

Facts About the World's Oceans

Ocean	Area (square mi)	Greatest Depth (ft)
Arctic	5,108,132	18,456
Atlantic	33,424,006	30,246
Indian	28,351,484	24,460
Pacific	64,185,629	35,837

1. If the depths of all the oceans were rounded to the nearest ten thousand, which two oceans would have the same depth?

_____

2. In 1960, scientists observed sea creatures living as far down as thirty thousand feet. In which ocean(s) could these creatures have lived?

_____

3. If you wanted to compare the depths of the Pacific Ocean and the Atlantic Ocean, which place value would you use to estimate?

_____

4. The oceans cover about three-fourths of Earth's surface. Estimate the total area of all the oceans combined by rounding to the nearest million.

_____

Choose the letter for the best answer.

5. There are 5,280 feet in a mile. About how many miles deep is the deepest point in the Pacific Ocean?

   A about 0.7 mile   C about 70 miles

   B about 7 miles   D about 700 miles

6. Rounding to the greatest place value, about how much larger is the Indian Ocean than the Arctic Ocean?

   F about 5 million sq. mi

   G about 10 million sq. mi

   H about 15 million sq. mi

   J about 25 million sq. mi

7. The Atlantic Ocean is about 40 times larger than the world's largest island, Greenland. Use this information to estimate the area of Greenland.

   A about 800,000 sq. mi

   B about 8,000,000 sq. mi

   C about 80,000,000 sq. mi

   D about 1,200,000,000 sq. mi

8. About how much larger would the Pacific Ocean have to be to have more area than the other three oceans combined?

   F about 2 hundred sq. mi

   G about 2 thousand sq. mi

   H about 2 million sq. mi

   J about 20 million sq. mi

Holt Mathematics

Reading Strategies

Draw Conclusions

In daily situations that involve math problems, an estimate is sometimes used rather than an exact answer. An **estimate** is an answer that **is close to the exact number.** Read these statements that give estimates:

- Over 45,000 fans attended the opening baseball game.
- The cost of admission is about $10.
- According to the map, we must drive about 50 miles.

In some situations, it is better to **overestimate.** Examples:
- the amount of money to take to the baseball game
- the driving time to the game

In these situations, an overestimate is best. This ensures that you have enough money and arrive at the game on time.

In other situations an **underestimate** would be best. Examples:
- the weight the ballpark express elevator can hold
- the number of "standing room only" tickets available

In these situations, an underestimate is best. This ensures that the elevator is not too heavy and that the "standing room only" section is not too crowded.

Tell whether an overestimate or an underestimate is best for each situation and why.

1. The weight that an airplane can hold.

_____

2. The amount of money for a trip.

_____

3. The number of people at a track meet.

_____

4. The number of people who can sit in a section of bleachers.

_____

5. The number of hours to drive from Chicago to New York.

_____

_____

Holt Mathematics

Name _____ Date _____ Class _____

LESSON 1-2

Puzzles, Twisters & Teasers

The Great Race

Sam and Lloyd are bicycle racing around Europe starting in Lisbon, Portugal. The race rules state that Sam and Lloyd must each travel a different route between cities. The final distances between cities for both Sam and Lloyd are listed below.

1. For each city, round the distance for each biker to the nearest 100.

2. Add the distance to the bikers' totals.

3. The race winner is the biker who completes the race with the lowest total distance (rounded to the nearest 100 miles).

4. As an added bonus, at each city an alphabet letter accompanies each racer's total. Put the letter from the winning racer at each city into the blanks below to solve the riddle. For example, Sam traveled a total of 300 miles to Madrid and Lloyd traveled a total of 400 miles. So write the "a" from Sam's total box in the first blank in the riddle.

Origin City	Sam's Route Distance	Lloyd's Route Distance	Sam's Total	Lloyd's Total
Lisbon, Portugal	—	—	0	0
Madrid, Spain	345	380	a	o
Paris, France	742	795	p	r
London, England	350	140	t	e
Brussels, Belgium	251	249	e	a
Berlin, Germany	453	357	f	p
Vienna, Austria	461	449	k	p
Athens, Greece	710	854	e	g
Rome, Italy	538	587	a	i
Lisbon, Portugal	1,171	1,046	r	l

AND THE WINNER IS: _____!

Riddle: What do you call a gorilla with a banana?

An ____ ___ ____ with ____ ___ ___ ____ ___ ___!

Holt Mathematics

LESSON **Practice A**
1-3 *Exponents*

Name the base and the exponent for each of the following.

1. 7^2

base _____

exponent _____

2. 5^4

base _____

exponent _____

3. 6^8

base _____

exponent _____

4. 5^9

base _____

exponent _____

5. 10^7

base _____

exponent _____

6. 4^3

base _____

exponent _____

Write using exponents.

7. 4×4

_____

8. $2 \times 2 \times 2$

_____

9. 10×10

_____

10. $5 \times 5 \times 5 \times 5$

_____

11. $3 \times 3 \times 3 \times 3$

_____

12. $8 \times 8 \times 8 \times 8 \times 8$

_____

Write as repeated multiplication.

13. 6^2

_____

14. 5^3

_____

15. 10^3

_____

16. 9^4

_____

17. 2^5

_____

18. 3^6

_____

19. How many different ways can you use the digits 3 and 5 to write expressions in exponential form? What are the expressions?

_____

20. What do the following two expressions have in common?
"three to the second power" and "three squared"

_____

Holt Mathematics

LESSON 1-3 Practice B
Exponents

Write each expression in exponential form.

1. 9×9

2. $7 \times 7 \times 7$

3. $1 \times 1 \times 1 \times 1 \times 1$

_____

_____

_____

4. $5 \times 5 \times 5 \times 5$

5. $2 \times 2 \times 2 \times 2 \times 2 \times 2$

6. $10 \times 10 \times 10 \times 10$

_____

_____

_____

Find each value.

7. 6^2

8. 5^3

9. 10^3

10. 7^2

_____

_____

_____

_____

11. 2^5

12. 3^4

13. 25^1

14. 16^0

_____

_____

_____

_____

Compare. Write $<$, $>$, or $=$.

15. 8^0 ☐ 7^1

16. 10^2 ☐ 11^2

17. 8^2 ☐ 4^3

18. 3^4 ☐ 5^2

19. 2^5 ☐ 9^2

20. 6^2 ☐ 3^3

21. What whole number equals 25 when it is squared and 125 when it is cubed?

_____

22. Use exponents to write the number 81 three different ways.

_____

Holt Mathematics

Practice C
Exponents

Write each expression in exponential form.

1. $10 \times 10 \times 10 \times 10$ **2.** $7 \times 7 \times 7 \times 7 \times 7$ **3.** $4 \times 4 \times 4$

_____   _____   _____

Find each value.

4. 8^2 _____   **5.** 4^3 _____   **6.** 6^3 _____   **7.** 15^2 _____

8. 2^8 _____   **9.** 3^5 _____   **10.** 38^1 _____   **11.** 7^3 _____

Compare. Write $<$, $>$, or $=$.

12. 8^2 ☐ 4^3 **13.** 9^2 ☐ 5^2 **14.** 6^2 ☐ 3^4

15. 7^2 ☐ 2^4 **16.** 10^2 ☐ 100^1 **17.** 81^0 ☐ 9^2

18. $4^2 + 5$ ☐ $3^3 - 7$ **19.** $2^3 + 2$ ☐ $3^2 - 2$ **20.** $2^5 - 10$ ☐ $4^2 + 6$

21. If it takes Cell A 3 hours to produce two cells, how many cells will Cell A produce in 24 hours?

_____

22. Use exponents to complete the table.

Generation	Number of People	Exponent
Parents	2	2^1
Grandparents	4	2^2
Great Grandparents		
Great-Great Grandparents		
Great-Great-Great Grandparents		

Holt Mathematics

CHAPTER	**Reteach**
1-3	*Exponents*

You can write a number in exponential form to show repeated multiplication. A number written in exponential form has a base and an exponent. An exponent tells you how many times a number, called the base, is used as a factor.

$8^4 \leftarrow$ exponent

$\uparrow$
base

Write the expression in exponential form.
$6 \times 6 \times 6$
6 is used as a factor 3 times.
$6 \times 6 \times 6 = 6^3$

Write each expression in exponential form.

1. $8 \times 8 \times 8 \times 8 \times 8$ **2.** 3×3 **3.** $5 \times 5 \times 5 \times 5$ **4.** $7 \times 7 \times 7$

_____   _____   _____   _____

You can find the value of expressions in exponential form.
Find the value.
2^5

Step 1: Write the expression as repeated multiplication.
$2^5 = 2 \times 2 \times 2 \times 2 \times 2$

Step 2: Multiply.
$2 \times 2 \times 2 \times 2 \times 2 = 32$

$2^5 = 32$

Find each value.

5. 12^3 **6.** 6^5 **7.** 10^4 **8.** 4^6

_____   _____   _____   _____

Holt Mathematics

Challenge
Exponent Riddle

What is the greatest number that can be written with two digits?

Find the value of each expression below. Then in the box at the bottom of the page, write each expression's letter in the blank above its value. When you have found all the values, you will have solved the riddle.

E 3^3 _____

H 5^2 _____

I 2^4 _____

N 34^0 _____

O 9^2 _____

P 4^3 _____

R 6^2 _____

T 7^2 _____

W 10^2 _____

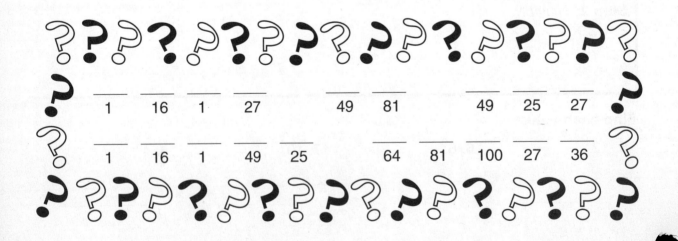

____ ____ ____ ____   ____ ____   ____ ____ ____
1   16   1   27   49   81   49   25   27

____ ____ ____ ____ ____   ____ ____ ____ ____ ____
1   16   1   49   25   64   81   100   27   36

Holt Mathematics

Problem Solving
LESSON 1-3 *Exponents*

1. The Sun is the center of our solar system. The Sun is the star closest to our planet. The surface temperature of the Sun is close to 10,000°F. Write 10,000 using exponents.

_____

2. Patty Berg has won 4^2 major women's titles in golf. Write 4^2 in standard form.

_____

3. William has 3^3 baseball cards and 4^3 football cards. Write the number of baseball cards and footballs cards that William has.

_____

4. Michelle recorded the number of miles she ran each day last year. She used the following expression to represent the total number of miles: $3 \times 3 \times 3 \times 3 \times 3 \times 3 \times 3$. Write this expression using exponents. How many miles did Michelle run last year?

_____

Choose the letter for the best answer.

5. In Tyrone's science class he is studying cells. Cell A divides every 30 minutes. If Tyrone starts with two cells, how many cells will he have in 3 hours?

   A 6 cells

   B 32 cells

   C 128 cells

   D 512 cells

6. Tanisha's soccer team has a phone tree in case a soccer game is postponed or cancelled. The coach calls 2 families. Then each family calls 2 other families. How many families will be notified during the 4th round of calls?

   F 2 families

   G 4 families

   H 8 families

   J 16 families

7. The Akashi-Kaiko Bridge is the longest suspension bridge in the world. It is located in Kobe-Naruto, Japan and was completed in 1998. It is about 3^8 feet long. Write the approximate length of the Akashi-Kaiko Bridge in standard form.

   A 6,561 feet

   B 2,187 feet

   C 512 feet

   D 24 feet

8. The Strahov Stadium is the largest sports stadium in the world. It is located in Prague, Czech Republic. Its capacity is about 12^5 people. Write the capacity of the Strahov Stadium in standard form.

   F 60 people

   G 144 people

   H 20,736 people

   J 248,832 people

25

Holt Mathematics

Reading Strategies

LESSON 1-3 *Synthesize Information*

Exponents are an efficient way to write repeated multiplication.

Read 2^4 ➞ *2 to the fourth power*

2^4 means **2 is a factor 4 times,** or $2 \times 2 \times 2 \times 2$

Read $2^4 = 16$ ➞ *2 to the fourth power equals 16.*

Exponent	Meaning	Value
10^3 *10 to the third power*	10 is a factor 3 times: $10 \times 10 \times 10$	$10^3 = 1,000$
6^5 *6 to the fifth power*	6 is a factor 5 times: $6 \times 6 \times 6 \times 6 \times 6$	$6^5 = 7,776$

Answer each question.

1. Write in words how you would read 3^4. _____

2. What does 3^4 mean? _____

3. What is the value of 3^4? _____

4. Write in words how you would read 5^3. _____

5. Write 5^3 as repeated multiplication. _____

6. What is the value of 5^3? _____

7. Tell why 2^3 is not 2 x 3.

_____

8. Is 3^4 the same as 4^3? _____ Explain why or why not.

_____

Holt Mathematics

Puzzles, Twisters & Teasers
LESSON 1-3
Answer This!

What are the only land mammals that cannot jump?

To find the answer:

1. Use a ruler to match each number and its value.
   (Each line you draw will cross a number and a letter)

2. Write the letter under the matching number in the decoder.

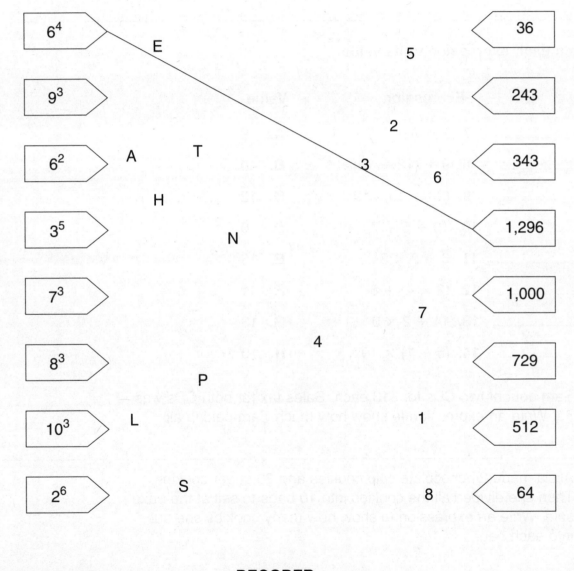

DECODER

3   4   3   1   2   5   6   7   8

___ ___ ___ ___ ___ ___ ___ ___ ___

Holt Mathematics

Practice A
Order of Operations

Name the operation you should perform first.

1. $5 + 6 \times 2$

2. $18 \div 3 - 1$

3. $4 + (7 - 1)$

_____

_____

_____

4. $3^2 + 6$

5. $(15 + 38) \times 6$

6. $5 \times 10 - 12$

_____

_____

_____

Match each expression to its value.

	Expression	Value
_____	**7.** $7 + 8 - 2$	**A.** 9
_____	**8.** $9 + (12 - 10)$	**B.** 40
_____	**9.** $(20 - 15) \times 2$	**C.** 12
_____	**10.** $10 \div 5 + 7$	**D.** 0
_____	**11.** $6 + 2 \times 3$	**E.** 16
_____	**12.** $(2 \times 4) + 8$	**F.** 11
_____	**13.** $14 \div 2 \times 0$	**G.** 13
_____	**14.** $(5 - 1) \times 10$	**H.** 10

15. Sam bought two CDs for $13 each. Sales tax for both CDs was
$3. Write an expression to show how much Sam paid in all.

_____

16. Alicia made 24 chocolate chip cookies and 36 sugar cookies.
Then she divided all the cookies into 10 bags to sell at the bake
sale. Write an expression to show how many cookies she put
into each bag.

_____

Holt Mathematics

Practice B
Order of Operations

Evaluate each expression.

1. $10 + 6 \times 2$

2. $(15 + 39) \div 6$

3. $(20 - 15) \times 2 + 1$

_____ _____ _____

4. $(4^2 + 6) \div 11$

5. $9 + (7 - 1) \times 2$

6. $(2 \times 4) + 8 - (5 \times 3)$

_____ _____ _____

7. $5 + 18 \div 3^2 - 1$

8. $8 + 5 \times 10 - 12$

9. $14 + (50 - 7^2) \times 3$

_____ _____ _____

Add parentheses so that each equation is correct.

10. $7 + 9 \times 3 - 1 = 25$

11. $2^3 - 7 \times 4 = 4$

12. $5 + 6 \times 9 \div 3 = 23$

_____ _____ _____

13. $12 \div 3 \times 2 = 2$

14. $8 + 3 \times 6 - 4 - 1 = 13$ **15.** $4 \times 3^2 + 1 = 40$

_____ _____ _____

16. $9 \times 0 + 5 - 3 = 42$

17. $15 \times 3^2 - 2^3 = 15$

18. $14 \div 2 + 5 \times 5 = 10$

_____ _____ _____

19. Tyler walked 2 miles a day for the first week of his exercise plan.
Then he walked 3 miles a day for the next 9 days. How many
miles did Tyler walk in all?

_____

20. Paulo's father bought 8 pizzas and 12 bottles of juice for the
class party. Each pizza cost $9 and each bottle of juice cost $2.
Paulo's father paid with a $100-bill. How much change did he
get back?

_____

Holt Mathematics

Practice C
LESSON 1-4
Order of Operations

Evaluate each expression.

1. $42 - 3 \times 10 + 2$

2. $1 + 4^3 - 16$

3. $(15 - 6) \times 2 + 20$

_____   _____   _____

4. $(5^2 + 3^2 + 2) \div 6$

5. $61 - 5 \times 2^3 + 5$

6. $7 \times 8 + (2 \times 4) \div 2^2$

_____   _____   _____

Add parentheses so that each equation is correct.

7. $12 - 3 \times 2 + 4^2 = 34$

8. $72 \div 2 \times 4 \div 3 = 3$

_____   _____

9. $13 + 7 - 6 + 4 \times 2 = 0$

10. $28 \div 7 + 3^3 - 3^2 - 1 = 21$

_____   _____

Use each of the numbers 2, 3, 4, and 6 once to make each equation correct.

11. $(\underline{\quad} - \underline{\quad}) + \underline{\quad} \times \underline{\quad} = 11$ **12.** $\underline{\quad} \times \underline{\quad} - (\underline{\quad} \div \underline{\quad}) = 6$

_____   _____

13. $\underline{\quad} + (\underline{\quad} \times \underline{\quad}) \times \underline{\quad} = 30$ **14.** $\underline{\quad} \div \underline{\quad} + \underline{\quad} \times \underline{\quad} = 20$

_____   _____

15. Use an exponent to write an expression with five 3s that has a value of 0.

_____

16. Mrs. Thompson is putting new tile on her bathroom floor. Each tile measures 2 inches on each side. The bathroom floor is 3 feet long and 2 feet wide. How many tiles will she use to cover the entire floor?

_____

Holt Mathematics

Reteach
LESSON 1-4 Order of Operations

A mathematical phrase that includes only numbers and operations is called a numerical expression.

$9 + 8 \times 3 \div 6$ is a numerical expression.

To evaluate a numerical expression, you find its value.

You can use the order of operations to evaluate a numerical expression.

Order of Operations
1. Do all operations within parentheses.
2. Find the values of numbers with exponents.
3. Multiply and divide in order from left to right.
4. Add and subtract in order from left to right.

Evaluate the expression.

$60 \div (7 + 3) + 3^2$	
$60 \div 10 + 3^2$	Do all operations within parentheses.
$60 \div 10 + 9$	Find the values of numbers with exponents.
$6 + 9$	Multiply and divide in order from left to right.
15	Add and subtract in order from left to right.

Evaluate each expression.

1. $7 \times (12 + 8) - 6$

$7 \times$ _____ $- 6$

_____ $- 6$

_____

2. $10 \times (12 + 34) + 3$

$10 \times$ _____ $+ 3$

_____ $+ 3$

_____

3. $10 + (6 \times 5) - 7$

$10 +$ _____ $- 7$

_____ $- 7$

_____

4. $2^3 + (10 - 4)$

_____

5. $7 + 3 \times (8 + 5)$

_____

6. $36 \div 4 + 11 \times 8$

_____

7. $5^2 - (2 \times 8) + 9$

_____

8. $3 \times (12 \div 4) - 2^2$

_____

9. $(3^3 + 10) - 2$

_____

Holt Mathematics

Challenge

Crack the Expression Code

Each of these symbols stands for a different operation symbol:

Each of these animals stands for a different whole number 1–4:

Use the equations below to find what each symbol and animal represents in the expression code.

 = 7

      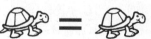

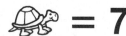

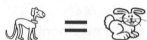

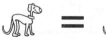

OPERATIONS		NUMBERS	
1. ♥ = _____		5. = _____	
2. ♠ = _____		6. = _____	
3. ♦ = _____		7. = _____	
4. ♣ = _____		8. = _____	

Holt Mathematics

LESSON 1-4

Problem Solving

Order of Operations

Evaluate each expression to complete the table.

Mammals with the Longest Tails

	Mammal	Expression	Tail Length
1.	Asian elephant	$2 + 3^2 \times 7 - (10 - 4)$	
2.	Leopard	$5 \times 6 + 5^2$	
3.	African elephant	$6 \times (72 \div 8) - 3$	
4.	African buffalo	$51 + 6^2 \div 9 - 12$	
5.	Giraffe	$4^3 - 3 \times 7$	
6.	Red kangaroo	$11 + 48 \div 6 \times 4$	

Choose the letter for the best answer.

7. Adam and his two brothers went to the zoo. Each ticket to enter the zoo costs $7. Adam bought two bags of peanuts for $4 each, and one of his brothers bought a lion poster for $12. Which expression shows how much money they spent at the zoo in all?

A $7 + 4 + 12$

B $7 \times 3 + 4 + 12$

C $7 \times 3 + 4 \times 2 + 12$

D $(7 \times 3) + (4 \times 12)$

8. An elephant eats about 500 pounds of grass and leaves every day. There are 2 Africa elephants and 3 Asian elephants living in the City Zoo. How many pounds of grass and leaves do the zookeepers need to order each week to feed all the elephants?

F 2,500 pounds

G 17,500 pounds

H 3,000 pounds

J 21,000 pounds

9. The average giraffe is 18 feet tall. Which of these expressions shows the height of a giraffe?

A $4^2 - 2$

B $3 \times 12 \div 4 + 2$

C $3^3 \div 9 \times 6$

D $20 \div 5 + 5 - 6$

10. Some kangaroos can cover 30 feet in a single jump! If a kangaroo could jump like that 150 times in a row, how much farther would it need to go to cover a mile? (1 mile = 5,280 feet)

F 780 feet   H 176 feet

G 26 feet   J 5,100 feet

Holt Mathematics

Name _____ Date _____ Class _____

Reading Strategies

Use a Flowchart

When you read a sentence, you read each word in order from left to right. To evaluate an expression, you cannot always compute the operations in the order they are given, from left to right. You must follow the order of operations. The order is listed in the flowchart below:

Parentheses	→	Exponents	→	Multiply or divide from left to right	→	Add or subtract from left to right

Example $3 \times (5 + 4) - 2^2$ **Parentheses**

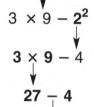

$3 \times 9 - 2^2$ **Exponents**

$3 \times 9 - 4$ **Multiply or divide (left to right)**

$27 - 4$ **Add or subtract (left to right)**

23

Answer each question.

1. Write a flowchart to list the order in which you would compute the operations for this expression: $12 - 2 \times 3 + 8 \div 2$.

_____

2. Evaluate this expression: $12 - 2 \times 3 + 8 \div 2$.

_____

3. Write a flowchart to list the order in which you would compute the operations for this expression: $(12 - 2) \times 3 + 8 \div 2$.

_____

4. Evaluate this expression: $(12 - 2) \times 3 + 8 \div 2$.

_____

5. Explain how Exercises 1 and 3 are alike and how they are different.

_____

_____

_____

Holt Mathematics

Name _____ Date _____ Class _____

Puzzles, Twisters & Teasers

Are You in Order?

What did one telephone say to the other when it proposed?

To answer the riddle, solve the following problems. Then write
the letter that is represented by each answer in the blanks below.

1. $20 + 16 \times 2 =$ _____ (I)

2. $55 \div (11 - 6) \times 8 =$ _____ (W)

3. $4 + 9 - (2 + 6) + 3 =$ _____ (L)

4. $(24 + 12) \div 12 =$ _____ (G)

5. $(4 + 6 \div 2) \times (1 + 9) =$ _____ (V)

6. $10 \times (54 - 49) + 17 =$ _____ (E)

7. $(36 \div 18)^3 + 17 \times 3 =$ _____ (Y)

8. $2^4 + (81 - 50) + 52 =$ _____ (O)

9. $21 \div (2 + 1) \times 5 - 2^2 =$ _____ (U)

10. $6 \div (1 + 2) \times 5^2 - 25 =$ _____ (A)

11. $32 \times (3 + 2) + 8 \div 2 =$ _____ (R)

12. $(6^3 \div 3) + 8 \div 2 =$ _____ (N)

I		___	___	___	___
		88	52	8	8

___   ___   ___   ___   ___   ___   ___
 3   52   70   67   59   99   31

___   ___   ___   ___   ___
25   164   52   76   3

Holt Mathematics

Name _____ Date _____ Class _____

Practice A

Mental Math

Choose the letter of the equation that shows the given property.

1. Associative Property

  A $2 + 3 = 3 + 2$

  B $7 \times 8 = 7 \times (4 + 4)$

  C $8 \times (6 \times 5) = (8 \times 6) \times 5$

  D $9 \times (2 + 4) = (9 \times 2) + (9 \times 4)$

2. Distributive Property

  F $3 \times (6 \times 11) = (3 \times 6) \times 11$

  G $75 + 15 = 15 + 75$

  H $9 \times 8 = 8 \times 9$

  J $12 \times (4 + 7) = (12 \times 4) + (12 \times 7)$

3. Commutative Property

  A $3 \times (7 + 8) = 3 \times 15$

  B $(10 + 4) + 3 = 10 + (4 + 3)$

  C $(9 + 2) \times 5 = (9 \times 5) + (2 \times 5)$

  D $6 \times 5 = 5 \times 6$

4. Associative Property

  F $20 \times (3 + 3) = (20 \times 3) + (20 \times 3)$

  G $4 + (3 + 9) = (4 + 3) + 9$

  H $(10 + 5) \times 7 = 15 \times 7$

  J $16 \times 8 = 8 \times 16$

Rewrite each expression using the named property.

5. $8 + 12$; Commutative Property

_____

6. $(9 \times 6) \times 4$; Associative Property

_____

7. $3 \times (5 + 2)$; Distributive Property

_____

8. $2 \times (4 + 5)$; Distributive Property

_____

Find each sum or product.

9. $7 + 15 + 3 + 5$

10. $7 \times 2 \times 5$

11. $4 \times 3 \times 5$

Multiply using the Distributive Property.

12. 4×38

13. 6×53

14. 8×42

15. Sue has \$4, Tom has \$11, Brian has \$6, and Anita has \$9. Use mental math to find how much money they have altogether.

_____

16. Each minibus seats 14 people, and the school owns 5 minibuses. Use mental math to find how many students can ride in the school's minibuses at the same time.

_____

Holt Mathematics

LESSON 1-5 **Practice B**
Mental Math

Evaluate.

1. 17 + 4 × 5

2. 25 × 3 × 4

3. 28 + 39 + 11 + 22

_____

_____

_____

4. 12 + 7 + 8 + 13

5. 10 + 3 × 2

6. 9 × 8 × 5

_____

_____

_____

7. 97 + 4 + 3 + 26

8. 2 × 6 × 5

9. 28 + 2 × 6

_____

_____

_____

Use the Distributive Property to find each product.

10. 4 × 16

11. 8 × 31

12. 3 × 62

13. 2 × 46

_____

_____

_____

_____

14. 5 × 29

15. 7 × 22

16. 9 × 21

17. 6 × 15

_____

_____

_____

_____

18. 8 × 44

19. 4 × 29

20. 7 × 31

21. 5 × 57

_____

_____

_____

_____

22. Each ticket to a play costs $27. How much will it cost to buy
4 tickets? Which property did you use to solve this problem
with mental math?

_____

23. Mr. Stanley bought two cases of pencils. Each case has 20 boxes.
In each box there is 10 pencils. Use mental math to find how
many pencils Mr. Stanley bought.

_____

24. When you consider that cows eat grass and the water needed to
grow the grass that cows eat, it takes 65 gallons of water to
produce one serving of milk! Use mental math to find how many
gallons of water are needed to produce 5 servings of milk.

_____

Holt Mathematics

LESSON 1-5 Practice C
Mental Math

Let *a*, *b*, and *c* represent three different numbers. Use them to write equations showing each property.

1. Associative Property

_____

2. Commutative Property

_____

3. Distributive Property

_____

Use mental math to find each sum or product.

4. $59 + 27 + 21 + 43$

_____

5. $4 \times 5 \times 8 \times 5$

_____

6. $25 \times 6 \times 5 \times 4$

_____

7. $175 + 318 + 82 + 25$

_____

8. $163 + 55 + 37$

_____

9. $5 \times 23 \times 6$

_____

Use the Distributive Property to find each product.

10. 19×7

_____

11. 52×40

_____

12. 62×5

_____

13. 32×9

_____

14. 11×15

_____

15. 7×37

_____

16. 4×108

_____

17. 84×20

_____

18. The Cineplex has 13 theaters. Four of the theaters seat 150 people, 8 of the theaters seat 250 people, and the largest theater seats 275 people. Use mental math to find how many people can see a movie at the Cineplex at one time.

_____

19. Antoine has two part-time jobs. He earns $17 an hour working construction during the day and $9 an hour stocking shelves in a hardware store at night. Last week he worked 20 hours at his construction job and 15 hours at the hardware store. Use mental math to find how much Antoine earned last week in all.

_____

Holt Mathematics

Reteach

LESSON 1-5

Mental Math

Commutative Property

Changing the order of addends does not change the sum.

$21 + 13 = 13 + 21$

Changing the order of factors does not change the product.

$5 \times 7 = 7 \times 5$

Associative Property

Changing the grouping of addends does not change the sum.

$(3 + 8) + 4 = 3 + (8 + 4)$

Changing the grouping of factors does not change the product.

$2 \times (7 \times 4) = (2 \times 7) \times 4$

Distributive Property

When you multiply a number by a sum, you can

• Find the sum and then multiply. $3 \times (8 + 4) = 3 \times 12 = 36$

or

• Multiply the number by each addend and then find the sum.

$3 \times (8 + 4) = (3 \times 8) + (3 \times 4) = 24 + 12 = 36$

Identify the property shown.

1. $3 \times (2 \times 6) = (3 \times 2) \times 6$

_____

2. $7 + 18 = 18 + 7$

_____

3. $4 \times (8 + 5) = 4 \times 13$

_____

4. $11 \times 8 = 8 \times 11$

_____

5. $3 \times (8 + 4) = (3 \times 8) + (3 \times 4)$

_____

6. $(3 + 8) + 4 = 3 + (8 + 4)$

_____

Identify the property shown and the missing number in each equation.

7. $9 + 16 = y + 9$

_____

8. $4 \times (3 \times 2) = (4 \times n) \times 2$

_____

9. $3 \times (11 + 4) = 3 \times a$

_____

10. $6 \times (9 + 14) = b \times 23$

_____

Holt Mathematics

Reteach

1-5 Mental Math (continued)

Find each sum or product.

A. $8 + 9 + 22 + 31$

$\quad 8 + 22 + 9 + 31$ Use the Commutative Property.

$\quad (8 + 22) + (9 + 31)$ Use the Associative Property.

$\quad\quad 30 \quad\quad + \quad 40$ Use mental math to add.

$\quad\quad\quad\quad 70$

B. $5 \times 7 \times 4$

$\quad 7 \times 5 \times 4$ Use the Commutative Property.

$\quad 7 \times (5 \times 4)$ Use the Associative Property.

$\quad 7 \times 20$ Use mental math to multiply.

$\quad 140$

Find each sum or product.

11. $3 + 58 + 27 + 22$ **12.** $8 \times 3 \times 5$ **13.** $5 \times 3 \times 4$

_____   _____   _____

14. $54 + 32 + 78 + 106$ **15.** $84 + 11 + 26 + 39$ **16.** $10 \times 3 \times 7$

_____   _____   _____

Find the product.

6×34

Step 1: Write one factor as a sum of two numbers.

$\quad 6 \times 34 = 6 \times (30 + 4)$

Step 2: Use the Distributive Property.

$\quad 6 \times (30 + 4) = (6 \times 30) + (6 \times 4)$

Step 3: Use mental math to multiply and add.

$\quad (6 \times 30) + (6 \times 4) = 180 + 24 = 204$

Use the Distributive Property to find each product.

17. 6×43 **18.** 12×34 **19.** 53×4 **20.** 74×8

_____   _____   _____   _____

Holt Mathematics

LESSON 1-5 Challenge
Magic Squares

When you add the numbers in each row, each column, and each diagonal of a magic square, you get the same number—the magic sum.

In the magic square at the right, for example, the magic sum is 30.

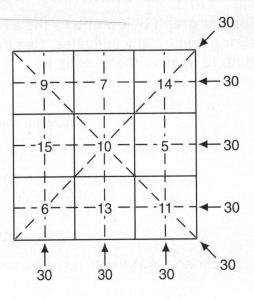

Use mental math to complete each magic square and find the magic sum.

1.

9	6	3
2		

Magic sum: _____

2.

11	4	
		8
7		

Magic sum: _____

3.

			4
12	6	7	9
			5
13	3		16

Magic sum: _____

4. Use mental math to create your own magic square using the numbers 1–9.

Magic sum: _____

41

Holt Mathematics

Name _____ Date _____ Class _____

Problem Solving
Mental Math

The bar graph below shows the average amounts of water used
during some daily activities. Use the bar graph and mental
math to answer the questions.

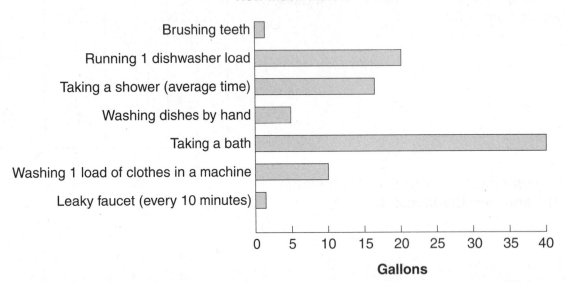

How Much Water?

1. Most people brush their teeth three
   times a day. How much water do
   they use for this activity every week?

_____

2. How much water is wasted in a day
   by a leaky faucet?

_____

3. The average American uses 124 gallons of water a day. Name
   a combination of activities listed in the table that would equal
   that daily total.

_____

_____

Choose the letter for the best answer.

4. Kenya used 24 gallons of water doing three of the activities listed in the table once.
   Which activities did she do?

   A taking a bath, brushing teeth, washing dishes by hand

   B taking a bath, brushing teeth, running 1 dishwasher load

   C taking a shower, brushing teeth, washing dishes by hand

   D taking a shower, brushing teeth, running 1 dishwasher load

5. If you wash two loads of dishes by hand instead of using a dishwasher, how much
   water do you save?

   F 30 gallons   **G** 15 gallons   **H** 10 gallons   **J** 1 gallon

Holt Mathematics

Reading Strategies
1-5 Focus on Vocabulary

The Commutative, Associative, and Distributive Properties of mathematics can make it easier to use mental math.

Commutative Property—The word **commute** means **to exchange.** In mathematics, when **addends or factors exchange places,** the sum or product is not affected.

Addends change places	Factors change places
13 + 18 + 17	4 × 7 × 5
13 + 17 + 18	4 × 5 × 7
30 + 18 = 48	20 × 7 = 140

Associative Property—The word **associate** means **to join.** In mathematics, **when addends or factors are joined, or grouped, with parentheses** in different ways, the sum or product is not affected.

Addends are grouped	Factors are grouped
11 + 4 + 16	7 × 8 × 5
11 + (4 + 16)	7 × (8 × 5)
11 + 20 = 31	7 × 40 = 280

Distributive Property—The word **distribute** means **to give out.** In mathematics, you can **distribute a factor** over a sum without affecting the original product.

5 × 17	17 = 10 + 7
(5 × 10) + (5 × 7)	Distribute 5 as a factor.
50 + 35	Multiply.
85	Add.

Answer each question.

1. Rewrite 17 + 8 + 13 using the Commutative Property, then compute.

_____

2. Rewrite 9 x 8 x 5 using the Associative Property, then compute.

_____

3. Rewrite 7 × 28 using the Distributive Property, then compute.

_____

43

Holt Mathematics

Puzzles, Twisters & Teasers

LESSON 1-5   *Who Is the Famous Movie Star?*

Use the Associative Property to find each sum or product.
Remember: $6 + 13 + 7 = 6 + (13 + 7)$ $18 \times 5 \times 2 = 18 \times (5 \times 2)$
$= 6 + 20 = 26$ $= 18 \times 10 = 180$

1. $7 + 6 + 4 =$ _____

2. $18 + 5 + 5 =$ _____

3. $17 + 3 + 10 =$ _____

4. $14 + 6 + 12 =$ _____

5. $4 \times 6 \times 10 =$ _____

6. $23 + 19 + 11 =$ _____

7. $39 \times 5 \times 2 =$ _____

8. $40 \times 5 \times 3 =$ _____

Use the Distributive Property to find the product.
Remember: $3 \times 16 = 3 \times (10 + 6)$
$= (3 \times 10) + (3 \times 6)$
$= 30 + 18 = 48$

9. $7 \times 15 =$

$(7 \times 10) + (7 \times 5)$

_____

10. $35 \times 4 =$ _____

11. $56 \times 4 =$ _____

12. $43 \times 7 =$ _____

13. $52 \times 9 =$ _____

14. $71 \times 11 =$ _____

15. $98 \times 12 =$ _____

16. $222 \times 9 =$ _____

Use the answers to connect the dots to see a famous movie star. Connect the dots in order from the least number to the greatest number. Then connect the dot for the greatest number to the dot for the least number.

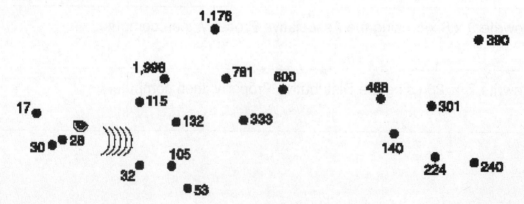

Holt Mathematics

LESSON 1-6 Practice A
Choose the Method of Computation

Answer the questions to describe the method of computation you should use to solve each problem.

PROBLEM: In the 2002 Winter Olympic Games, the United States won 10 gold medals, 13 silver medals, and 11 bronze medals. How many medals did the United States win in all?

1. What method of computation will you use to solve this problem?

_____

_____

2. Why did you choose this method of computation?

_____

_____

3. What is the solution to the problem?

_____

_____

PROBLEM: The United States holds the record for the most Summer Olympic Medals ever won. As of 2004, the United States had won 850 gold medals, 661 silver medals, and 563 bronze medals. How many Summer Olympic medals has the United States won in all?

4. What method of computation will you use to solve this problem?

_____

_____

5. Why did you choose this method of computation?

_____

_____

6. What is the solution to the problem?

_____

_____

Holt Mathematics

LESSON 1-6 Practice B
Choose the Method of Computation

1. Athletes from 197 countries competed at the 1996 Summer Olympic Games held in Atlanta, Georgia. That is 25 more countries that competed at the 1992 games held in Barcelona, Spain. How many different countries competed in Barcelona?

_____

2. At the 1996 Summer Olympic Games held in Atlanta, Georgia, 10,310 athletes competed. At the 1992 Summer Olympic Games held in Barcelona, Spain, 9,364 athletes competed. How many more athletes competed in Atlanta than in Barcelona?

_____

3. The marathon race is one of the oldest events in the Summer Olympic Games. Marathon competitors run a total of 26 miles 385 yards. There are 5,280 feet in a mile and 3 feet in a yard. How many yards long is the entire marathon race?

_____

4. The world record for the fastest men's marathon race is 2 hours, 5 minutes, 42 seconds. The world record for the fastest women's marathon race is 2 hours, 20 minutes, 43 seconds. How much faster is the men's record marathon time?

_____

5. The men's outdoor world record in the high jump is 2.45 meters or 8 feet 0.5 inches. The women's outdoor world record in the high jump is 2.09 meters or 6 feet 10.25 inches. How much higher is the men's high jump record? Write the answer in meters and feet.

_____

6. The men's world record in the 400-meter relay is 37.40 seconds, held by the U.S. If each of the four runners each ran 100 meters in the same time, how long did each runner run?

_____

7. Athletes from 13 nations competed in the first modern Olympics in 1896. Today, athletes from nearly 200 nations compete in the Summer Olympics. About how many more nations participate in the Olympics today than in 1896?

_____

Holt Mathematics

Name _____ Date _____ Class _____

Practice C

Choose the Method of Computation

Use the information below to answer the questions.

The Paralympics are Olympics Games held for the disabled. The
Paralympics are held by the Olympic host country in the same year
and usually the same city. The XII Paralympic Summer Games were
held in Athens, Greece in 2004. The countries with the most medals
are as follows; China 141 (63 gold), Australia 100 (26 gold), and
Great Britain 94 (35 gold). Almost 4,000 athletes from 136 nations
competed in 19 sports.

1. How many gold medals did the top three countries with the most
   medals win?

   _____

2. During the 2004 Summer Olympic Games, about 11,000
   athletes competed in 28 sports. About how many more athletes
   competed in the 2004 Summer Olympic Games than in the 2004
   Summer Paralympic Games?

   _____

3. How many more medals did China win than Australia?
   gold medals?

   _____

4. During the 2004 Summer Olympic Games, the U.S. won the
   most medals with 103 medals (35 gold). Compare the country
   with the most medals in the Summer Olympic Games with the
   country with the most medals in the Summer Paralympic
   Games. Which country had more medals? how many more?
   how many more gold medals?

   _____

5. During the 2004 Summer Olympic Games, China won 63
   medals (16 gold). How many medals in all did China take home
   in 2004? gold medals?

   _____

6. During the 2004 Summer Olympic Games, Great Britain won
   30 medals (9 gold). How many more medals in all did Great Britain
   win in the 2004 Summer Paralympic Games than in the Summer Olympic
   Games? gold medals?

   _____

Holt Mathematics

Reteach

1-6 *Choose the Method of Computation*

Paper and pencil, mental math, and a calculator are three computation methods for solving problems.

• If there are many small numbers, use paper and pencil.

• If the numbers are small and easy, use mental math.

• If the numbers are large, use a calculator.

Before you solve a problem decide which computation method is the best.

Choose a computation method. Then solve.

At a book fair, 76 books were sold on the first day and 82 books were sold on the second day. How many books were sold during the two days?

Number of books sold on the first day + Number of books sold on the second day

| 76 | + | 82 |

The numbers are small and 82 is close to a multiple of 10. You can use mental math.

(76 + 2) + (82 − 2) = 78 + 80 = 158

During the two days, 158 books were sold.

Choose a computation method. Then solve.

1. Of the 248 books on display, 46 were nonfiction books. How many books were not nonfiction books?

   _____

2. Lisa bought 2 biographies for $5.37 each, a novel for $7.95, and a bookmark for $1.19. How much money did Lisa spend?

   _____

3. Over two days, 234 students visited the book fair in groups of 18. How many groups visited the fair?

   _____

Holt Mathematics

Challenge
Finger Math

Chisenbop is an ancient method of computation using your fingers. One of the best-known forms of Chisenbop is used for basic multiplication computations. It works only when all the factors are greater than 5. Follow these steps to use this form of the Chisenbop method of computation. The product of 6 × 7 is shown as an example.

Step 1 Subtract 5 from the first factor. Turn down that number of fingers on your left hand.

Step 2 Subtract 5 from the second factor. Turn down that number of fingers on your right hand.

Step 3 Multiply the total number of turned-down fingers on both hands by 10.

Step 4 Find the product of the numbers of fingers that are **not** turned down on each hand.

Step 5 Add the two products from Step 3 and Step 4.

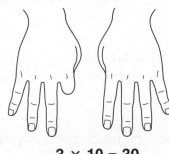

$3 \times 10 = 30$
↑
both hands

$4 \times 3 = 12$
↑   ↑
left hand   **right hand**

$30 + 12 = 42$
So, 6 × 7 = 42.

Use the Chisenbop method to find each product.

1. 7 × 8 = _____

2. 6 × 9 = _____

3. 8 × 6 = _____

4. 8 × 9 = _____

5. 6 × 6 = _____

6. 9 × 7 = _____

7. 7 × 7 = _____

8. 7 × 9 = _____

9. 8 × 8 = _____

10. 6 × 8 = _____

11. 9 × 9 = _____

12. 9 × 8 = _____

13. When would you choose to use the Chisenbop method of computation? When would you choose not to use that method? Explain.

_____

_____

_____

Holt Mathematics

Name _____ Date _____ Class _____

Use the table below to answer questions 1–6. For each question, write the method of computation you should use to solve it. Then write the solution.

1. How many bones are in an average person's arms and hands altogether?

_____

2. How many more bones are in an average person's head than chest?

_____

3. Which part of the body has twice as many bones as the spine?

_____

4. How many bones are in the body altogether?

Bones in the Human Body

Body Part	Number of Bones
Head	28
Throat	1
Spine	26
Chest	25
Shoulders	4
Arms	6
Hands	54
Legs	10
Feet	52

5. A newborn baby has 350 bones. How many more bones does a newborn baby have than an adult?

_____

_____

6. How many bones are in each of an average person's feet, hands, legs, and arms?

_____

_____

Choose the letter for the best answer.

7. The body's longest bones— thighbones and shinbones—are in the legs. The average thighbone is about 20 inches long, and the average shinbone is about 17 inches long. What is the total length of those four bones?

   A paper and pencil; 74 inches

   B paper and pencil; 37 inches

   C mental math; 20 inches

   D calculator; 17 inches

8. The body has 650 muscles. Seventeen of those muscles are used to smile and 42 muscles are used to frown. How many more muscles are used to frown than to smile?

   F mental math; 35 muscles

   G mental math; 25 muscles

   H paper and pencil; 608 muscles

   J calculator; 633 muscles

Holt Mathematics

LESSON 1-6

Reading Strategies
Analyze Information

Read each problem. Then use the four steps to help you solve each problem.

> Marta kept track of the points she earned on 5 math tests. They were 85, 76, 88, 78, and 91. The total number of points possible on all 5 tests is 500 points. How many points has Marta earned so far?

1. What question is asked?

_____

2. What information is needed to answer the question?

_____

3. Circle the operation needed to solve the problem.

• Addition   • Subtraction   • Multiplication   • Division

4. Show how you compute and solve the problem.

_____

> Marta spends 35 minutes each night on her math homework. She spends another 45 minutes each night on the rest of her homework. How much time does Marta spend studying math over 5 days?

5. What question is asked?

_____

6. What information is needed to answer the question?

_____

7. Circle the operation needed to solve the problem.

• Addition   • Subtraction   • Multiplication   • Division

8. Show how you compute and solve the problem.

_____

Holt Mathematics

LESSON 1-6

Puzzles, Twisters & Teasers

Crossword Mania

Complete the crossword puzzle.

Across

1. Good computation method to use to solve number 3 down.

3. The number of seconds in a week is how much greater than 604,780?

5. The number of months in a century is how many times greater than the number of years in a century?

6. 121,780 is how many times greater than 12,178?

8. You can use this machine and the software that comes with it to do complex calculations. It can even help people do their taxes.

9. The best calculating tool humans have; it helps you do mental math.

Down

2. Good computation method to solve number 3 across.

3. A cheetah can run 60 miles an hour. If it could maintain that speed for half an hour, how many miles could it run?

4. You can find the answer to number 5 across inside this.

7. 56 × 42 is how much less than 2,361?

Holt Mathematics

LESSON 1-7 Practice A
Patterns and Sequences

Choose the sequence that matches each pattern.

1. Start with 12; subtract 2.

  A 2, 4, 6, 8, 10, 12, …

  B 12, 11, 10, 9, 8, 7, …

  C 12, 14, 16, 18, 20, …

  D 12, 10, 8, 6, 4, 2, …

2. Start with 3; multiply by 2.

  F 3, 5, 7, 9, 11, …

  G 3, 6, 12, 24, 48, …

  H 2, 6, 18, 54, 162, …

  J 3, 4, 5, 6, 7, 8, 9, …

3. Start with 5; add 4.

  A 5, 4, 3, 2, 1, …

  B 5, 9, 13, 17, 21, …

  C 5, 10, 15, 20, 25 …

  D 4, 9, 14, 19, 24, …

4. Start with 1; multiply by 10.

  F 1, 10, 100, 1,000, …

  G 1, 10, 20, 30, 40, …

  H 10, 100, 1,000, 10,000, …

  J 10, 20, 30, 40, 50, …

Identify a pattern in each sequence.

5. 1, 4, 7, 10, 13, …   **6.** 15, 13, 11, 9, …   **7.** 5, 10, 15, 20, …

_____   _____   _____

8.

Position	1	2	3	4	5	6	7
Value of Term	10	20	30	40	50	60	70

_____

Name the next three terms in each sequence.

9. 1, 6, 11, 16, 21, □, □, □, …   **10.** 2, 4, 6, 8, 10, □, □, □, …

_____   _____

11.

Position	1	2	3	4	5	6	7
Value of Term	50	45	40	35	30	25	20

12. The temperature was 79°F on Friday, 76°F on Saturday, and 73°F on Sunday. If this weather pattern continues, what will the temperature be on Monday?

13. Tony's Cafe sells four sizes of pizza. The first three sizes are 8 inches, 10 inches, and 12 inches. If this pattern of size continues, what is the largest size pizza the cafe sells?

_____

Holt Mathematics

LESSON 1-7 Practice B
Patterns and Sequences

Identify a pattern in each sequence and then find the missing terms.

1. 4, 8, 16, 32, □, □, □, …

2. 100, 95, 90, 85, □, □, □, …

_____

3. 8, 20, 32, 44, □, □, □, …

4. 6, 12, 18, 24, □, □, □, …

_____

5. 9, 18, 27, 36, □, □, □, …

6. 3, 6, 12, 24, □, □, □, …

_____

7.

Position	1	2	3	4	5	6	7
Value of Term	5	10	20	40			

8. 300, 250, □, □, 100, □, 0, …

9. 1, 15, □, 43, 57, □, 85, 99, …

_____

10. 7, □, 21, 28, □, □, □, 56, …

11. 9, □, 13, □, □, □, 21, 23, …

_____

12.

Position	1	2	3	4	5	6	7
Value of Term	3	12	21	30			

13. A forest ranger in Australia took measurements of a eucalyptus tree for the past 3 weeks. The tree was 12 inches tall the first week, 19 inches the second week, and 26 inches the third week. If this growth pattern continues, how tall will the tree be next week?

_____

14. Maria puts the same amount of money in her savings account each month. She had $450 in the account in April, $600 in May, and $750 in June. If she continues her savings pattern, how much money will she have in the account in July?

_____

Holt Mathematics

Name _____ Date _____ Class _____

LESSON 1-7 Practice C
Patterns and Sequences

Identify a pattern in each sequence, and then find the missing terms.

1. 13, 17, 21, 25, ☐, ☐, ☐, …

_____

2. 48, 45, 42, 39, ☐, ☐, ☐, …

_____

3. 1,600, 800, 400, 200, ☐, ☐, ☐, …

_____

4. 1, 3, 9, 27, ☐, ☐, ☐, …

_____

5.

Position	1	2	3	4	5	6	7
Value of Term	19	38	57	76			

Identify a pattern in each sequence. Name the missing terms.

6. 2, ☐, ☐, 16, 32, ☐, …

_____

7. 8, ☐, 24, ☐, 40, ☐, 56, …

_____

8.

Position	1	2	3	4	5	6	7
Value of Term	4	12	14	42			

9. The sequence 1, 1, 2, 3, 5, 8, 13, … is called the Fibonacci Sequence. Identify the pattern in the sequence, and name the next three terms.

_____

_____

10. The pattern below shows a special sequence called triangular numbers. Draw the next figure in the pattern, and name the next triangular number in the sequence.

   1   3   6   _____

Holt Mathematics

LESSON 1-7 Reteach
Patterns and Sequences

Find the next three numbers in the sequence.

8, 12, 16, 20, 24, □, □, □, ...

Step 1: Look at pairs of numbers to find the pattern.

8, 12, 16, 20, 24, □, □, □, ...

8 + 4 = 12   12 + 4 = 16   16 + 4 = 20

The pattern is to add 4.

Step 2: Use the pattern to name the next three numbers.

24 + 4 = 28   28 + 4 = 32   32 + 4 = 36

The next three numbers are 28, 32, and 36.

Find the next three numbers in each sequence.

1. 5, 8, 6, 9, 7 □, □, □, ...

_____

2. 90, 80, 70, 60, □, □, □, ...

_____

3. 2, 8, 4, 16, □, □, □, ...

_____

4. 10, 14, 18, 22, □, □, □, ...

_____

5. 13, 21, 29, 37, □, □, □, ...

_____

6. 24, 12, 16, 8, □, □, □, ...

_____

7. 14, 12, 10, 8, □, □, □, ...

_____

8. 1, 7, 13, 19, □, □, □, ...

_____

9. 1, 3, 6, 10, □, □, □, ...

_____

10. 40, 38, 36, □, □, □, ...

_____

11. 54, 45, 36, □, □, □, ...

_____

12. 10, 25, 40, □, □, □, ...

_____

13. 36, 29, 22, □, □, □, ...

_____

14. 18, 36, 72, □, □, □, ...

_____

Holt Mathematics

Name _____ Date _____ Class _____

Draw the next three figures in each pattern.

1.
_____

2.
_____

3. 🌸 🌷 🌸 🌳 🌷 🌳 🌸 🌷 🌸
_____

4.
_____

5. 😊 😊 😊 👧 👧 👧 😊 😊 😊
_____

6.
_____

7.
_____

Holt Mathematics

Name _____ Date _____ Class _____

Problem Solving
Patterns and Sequences

1. A giant bamboo plant was 5 inches tall on Monday, 23 inches tall on Tuesday, 41 inches tall on Wednesday, and 59 inches tall on Thursday. Describe the pattern. If the pattern continues, how tall will the giant bamboo plant be on Friday, Saturday, and Sunday?

_____

_____

_____

2. A scientist was studying a cell. After the second hour there were two cells. After the third hour there were four cells. After the fourth hour there were eight cells. Describe the pattern. If the pattern continues, how many cells will there be after the fifth, sixth, and seventh hour?

_____

_____

_____

Choose the letter for the best answer.

3. The first place prize for a sweepstakes is $8,000. The third place prize is $2,000. The fourth place prize is $1,000. The fifth place prize is $500. What is the second place prize?

   A $7,000   **C** $4,000

   B $6,000   **D** $3,000

4. The temperature was 59°F at 3:00 A.M., 62°F at 5:00 A.M., and 65°F at 7:00 A.M. If the pattern continues, what will the temperature be at 9:00 A.M., 11:00 A.M., and 1:00 P.M.?

   F 66°F at 9:00 A.M., 67°F at 11:00 A.M., 68°F at 1:00 P.M.

   G 68°F at 9:00 A.M., 70°F at 11:00 A.M., 72°F at 1:00 P.M.

   H 68°F at 9:00 A.M., 71°F at 11:00 A.M., 74°F at 1:00 P.M.

   J 70°F at 9:00 A.M., 75°F at 11:00 A.M., 80°F at 1:00 P.M.

Holt Mathematics

Name _____ Date _____ Class _____

Reading Strategies
Sequence

When a story has several parts, such as *Star Wars,* you want to watch them in order, or in **sequence.** When a set of numbers is **in order,** it is called a **sequence.** A sequence of numbers may have a special pattern. When we talk about the pattern, each number in the sequence is called a **term.**

Read each term of the sequence below. Note how the numbers change from one term to the next. What do you think is the pattern in this sequence?

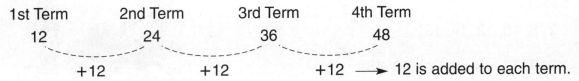

1st Term	2nd Term	3rd Term	4th Term
12	24	36	48

+12   +12   +12 ⟶ 12 is added to each term.

The pattern for this sequence of terms is + 12. More terms can be added to a sequence by continuing the pattern.

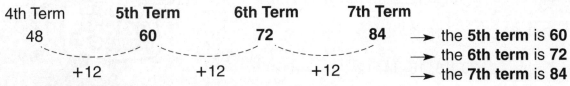

4th Term	**5th Term**	**6th Term**	**7th Term**
48	**60**	**72**	**84**

+12   +12   +12

⟶ the **5th term** is **60**
⟶ the **6th term** is **72**
⟶ the **7th term** is **84**

Use this sequence to answer Exercises 1–6:
   78   77   75   72   68

1. What is the 3rd term in the sequence? _____

2. What change occurs between the 1st term and 2nd term?

_____

3. What change occurs between the 2nd term and the 3rd term?

_____

4. What change occurs between the 3rd term and the 4th term?

_____

5. Write a rule to describe how to find each term in the sequence.

_____

_____

6. Identify the 6th, 7th, and 8th terms in this sequence.

_____

Holt Mathematics

Puzzles, Twisters & Teasers

LESSON 1-7

Have You Heard?

Did you hear about the student who tried using a broken calculator to evaluate $29^7 + 13^4 + 5^8$?

Complete each sequence. Use the decoder below to find the letter that corresponds to the answer for each exercise. Place each letter above it's corresponding exercise number below to complete the answer to the puzzle.

1. 3, 6, 9, 12, _____

2. 8, 10, 12, 14, 16, _____

3. 8, 16, _____, 32, 40, 48

4. _____, 10, 15, 20, 25, 30

5. 20, 16, 12, 8, 4, _____

6. 0, 6, _____, 18, 24, 30

7. _____, 20, 18, 16, 14, 12

8. 1, 2, 4, 8, 16, _____

9. 7, 14, 21, _____, 35

10. 3, _____, 12, 24, 48

0 = B	18 = S
5 = Y	22 = E
6 = C	24 = A
12 = O	28 = I
15 = N	32 = D

He couldn't even reach __ __ __ __ __ __
　　　　　　　　　　　2　7　10　6　1　8

__ __ __ __ .
5　3　2　7

Holt Mathematics

Name _____ Date _____ Class _____

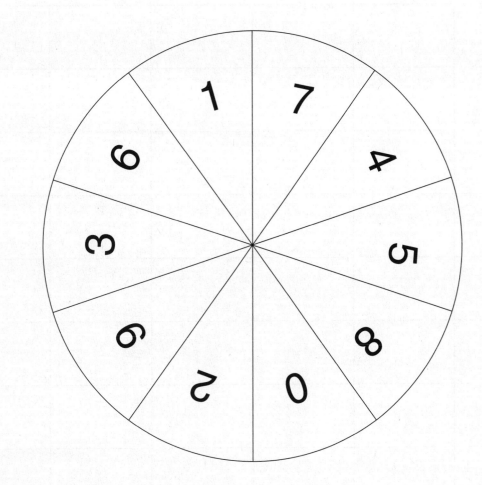

Name _____ Date _____ Class _____

Teacher Tool

Place Value Chart

Hundred thousands	Ten thousands	Thousands	Hundreds	Tens	Ones

Holt Mathematics

Practice A
Comparing and Ordering Whole Numbers

Write <, >, or = to compare the numbers.

1. 8 $<$ 18   **2.** 43 $>$ 34   **3.** 100 $>$ 90

4. 295 $>$ 259   **5.** 706 $=$ 706   **6.** 1,006 $<$ 6,001

Write the numbers from least to greatest.

7. 3; 13; 1   **8.** 88; 80; 78   **9.** 104; 204; 102

   1; 3; 13   78; 80; 88   102; 104; 204

10. 75; 95; 59   **11.** 642; 855; 658   **12.** 274; 207; 740

   59; 75; 95   642; 658; 855   207; 274; 740

Write the numbers from greatest to least.

13. 10; 100; 11   **14.** 36; 16; 63   **15.** 28; 20; 80

   100; 11; 10   63; 36; 16   80; 28; 20

16. 500; 300; 305   **17.** 593; 93; 59   **18.** 184; 800; 481

   500; 305; 300   593; 93; 59   800; 481; 184

19. English is spoken in 47 countries around the world. French is spoken in 23 countries. Which language is spoken in the most countries?

English

20. The United States–Mexico border is 1,933 miles long. The United States–Canada border is 3,987 miles long. Which border is longer?

the United States–Canada border

Holt Mathematics

Practice B
Comparing and Ordering Whole Numbers

Compare. Write <, >, or =.

1. 69 $<$ 96   **2.** 117 $>$ 107   **3.** 958 $<$ 9,124

4. 3,567 $=$ 3,567   **5.** 18,443 $>$ 1,844   **6.** 64,209 $<$ 64,290

Order the numbers from least to greatest.

7. 58; 166; 85   **8.** 115; 151; 111   **9.** 269; 29; 96

   58; 85; 166   111; 115; 151   29; 96; 269

10. 308; 3,800; 3,080   **11.** 1,864; 824; 1,648   **12.** 4,663; 4,336; 43,666

   308; 3,080; 3,800   824; 1,648; 1,864   4,336; 4,663; 43,666

Order the numbers from greatest to least.

13. 35; 53; 13   **14.** 807; 800; 708   **15.** 249; 392; 248

   53; 35; 13   807; 800; 708   392; 249; 248

16. 555; 600; 535   **17.** 7,320; 6,000; 6,305   **18.** 999; 9,559; 5,995

   600; 555; 535   7,320; 6,305; 6,000   9,559; 5,995; 999

19. Delaware and Rhode Island are the two smallest states. Delaware covers 1,955 square miles, and Rhode Island covers 1,045 square miles. What is the smallest state in the United States?

Rhode Island

20. Vermont and Wyoming have the smallest populations in the United States. The population of Vermont is 608,827. The population of Wyoming is 493,782. Which state has the smallest population?

Wyoming

Holt Mathematics

Practice C
Comparing and Ordering Whole Numbers

Compare. Write <, >, or =.

1. 1,478 $<$ 1,748   **2.** 5,643 $=$ 5,643

3. 9,610 $<$ 10,961   **4.** 308,524 $>$ 3,854

Order the numbers from least to greatest.

5. 379; 79; 978   **6.** 16,780; 17,847; 6,988

   79; 379; 978   6,988; 16,780; 17,847

7. 76,334; 47,961; 70,336   **8.** 101,695; 19,568; 191,658

   47,961; 70,336; 76,334   19,568; 101,695; 191,658

Order the numbers from greatest to least.

9. 605; 560; 565   **10.** 8,320; 8,063; 8,663

   605; 565; 560   8,663; 8,320; 8,063

11. 49,210; 49,000; 49,910   **12.** 352,699; 353,963; 95,614

   49,910; 49,210; 49,000   353,963; 352,699; 95,614

13. Alaska, California, and Texas are the three largest states. Alaska covers 615,230 square miles. California covers 158,869 square miles. Texas covers 267,277 square miles. Write the states in order by size, from largest to smallest.

Alaska, Texas, California

14. California, New York, and Texas have the largest populations in the United States. Their populations are 33,871,648; 20,851,820; and 18,976,457. California has the largest population. More people live in Texas than in New York. What is each state's population?

CA: 33,871,648; TX: 20,851,820; NY: 18,976,457

Holt Mathematics

Reteach
Comparing and Ordering Whole Numbers

You can use place value to compare or order whole numbers.
Use < or > to compare the numbers.

289,865 ☐ 289,765

Thousands			Ones			Compare the digits from left to right.
H	T	O	H	T	O	
2	8	9	8	6	5	First Number
2	8	9	7	6	5	Second Number

8 > 7
So, 289,865 > 289,765

Write < or > to compare the numbers.

1.

Thousands			Ones		
H	T	O	H	T	O
	3	5	4	7	
	3	5	3	2	

3,547 $>$ 3,532

2.

Thousands			Ones		
H	T	O	H	T	O
	9	5	3	6	
	9	6	3	5	

9,536 $<$ 9,635

Write the numbers in order from least to greatest.
976; 859; 924

Ones		
H	T	O
9	7	6
8	5	9
9	2	4

Compare the numbers in pairs.
976 > 859, 976 > 924, and 859 < 924.
So the numbers from least to greatest are 859; 924; 976.

Write the numbers in order from least to greatest.

3.

Ones		
9	5	4
9	4	5
9	6	9

945; 954; 969

4.

Ones		
3	4	3
3	3	4
4	3	4

334; 343; 434

5.

Ones		
8	9	4
8	9	2
9	6	5

892; 894; 965

Holt Mathematics

Holt Mathematics

Challenge
Ancient Calculators

Long before place-value charts were invented, people used a tool called an abacus to show and read numbers. The Chinese began using the *Suan Pan* abacus, shown below, about 800 years ago.

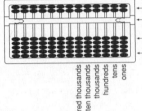

Each bead above the center bar stands for 5 units.

Center bar

Each bead below the center bar stands for 1 unit.

Each rod, or row, of beads represents one place value.

hundred thousands
ten thousands
thousands
hundreds
tens
ones

To use an abacus to show and read a number, move the beads to the center bar for each place value and add. For example:

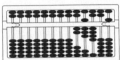 2 8 4 0 5 = 28,405   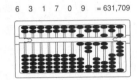 6 3 1 7 0 9 = 631,709

Write the number shown on each abacus.
Then write < or > to compare the numbers.

1. 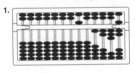   2.

501, 237 $>$ 492,668

7   **Holt Mathematics**

Problem Solving
Comparing and Ordering Whole Numbers

Use the tables below to answer each question.

Most Populated Countries	
Brazil	174,468,575
China	1,273,111,290
India	1,029,991,145
Indonesia	228,437,870
United States	278,058,881

Largest Countries (square mi)	
Brazil	3,265,059
Canada	3,849,646
China	3,705,408
Russia	6,592,812
United States	3,539,224

1. Which country has the greatest population?

   China; 1,273,111,290

2. Which countries have more than one billion people?

   China and India

3. Which country is the largest in the world?

   Russia

4. Which country's area is closest to 4,000,000 square miles?

   Canada

5. What is the error in the following statement? Canada is larger than the United States, but smaller than China.

   Canada is larger than China.

6. Based on population and size, which country do you think is more crowded, Brazil or the United States? Explain.

   U.S.; They are about the same size, but U.S. has more people.

7. Which country has a population less than two hundred million?
   A China   **C** Brazil
   B Indonesia   D India

8. Which countries have populations greater than the United States?
   F China and Brazil
   G China and India
   H India and Indonesia
   J Indonesia and China

9. Which list shows the countries in order by population from greatest to least?
   A China, United States, India, Indonesia, Brazil
   B China, India, Indonesia, Brazil, United States,
   C China, India, Indonesia, United States, Brazil
   D China, India, United States, Indonesia, Brazil

10. Which list shows the countries in order by size from smallest to largest?
    F Brazil, United States, China, Canada, Russia
    G Brazil, United States, Canada, China, Russia
    H Brazil, United States, Canada, Russia, China
    J Brazil, United States, Russia, China, Canada

8   **Holt Mathematics**

Reading Strategies
Analyze Information

Reading numbers helps you compare them.

Number	Read
2,581	2 thousand, 581
6,328	6 thousand, 328

Compare greatest place value: thousands.
6 thousand is greater than 2 thousand. So,

6,328 > 2,581 or 2,581 < 6,328.

Larger numbers can be compared and ordered in the same way.

Number	Read
453,276,328	453 million, 276 thousand, 328
435,617,119	435 million, 617 thousand, 119
457,428,937	457 million, 428 thousand, 937

Compare greatest place value: millions.
457 million > 453 million > 435 million

These three numbers in order from greatest to least are
457,428,937; 453,276,328; 435,617,119.

Use the numbers 637,598 and 673,522 to answer Exercises 1–3.

1. Write how these two numbers are read.

   637,598 _____ 637 thousand, 598
   673,522 _____ 673 thousand, 522

2. Which place value will you compare to decide which number is greater? _____ ten thousands

3. Use > or < to compare. 637,598 $<$ 673,522

Use the numbers 353,276,128; 353,268,437; and 353,248,753 to answer Exercises 4–6.

4. Write how these numbers are read.

   353,276,128 _____ 353 million, 276 thousand, 128
   353,268,437 _____ 353 million, 268 thousand, 437
   353,248,753 _____ 353 million, 248 thousand, 753

5. Which place value will you compare to put the numbers in order? _____ ten thousands

6. Order these three numbers from least to greatest.

   353,248,753; 353,268,437; 353,276,128

9   **Holt Mathematics**

Puzzles, Twisters & Teasers
Did you know?

What is one way tarantulas defend themselves?

Let's find out! Circle the greater number. Put the letter above the greater number in the boxes below.

	T	L		W	H
1.	(691)	619	11.	99,987	(100,000)
	S	H		U	A
2.	5,618	(8,567)	12.	5,009	(5,509)
	E	I		W	I
3.	(2,649)	2,469	13.	10,080	(10,800)
	S	Y		R	Q
4.	957	(975)	14.	(687)	678
	L	C		A	B
5.	(13,485)	12,635	15.	1,013	(1,173)
	A	I		E	A
6.	(873)	854	16.	213,946	(214,026)
	E	U		L	V
7.	512,009	(512,125)	17.	(1,217)	1,127
	T	N		M	L
8.	308	(380)	18.	917	(971)
	C	K		S	N
9.	(7,498)	7,398	19.	(2,913)	2,513
	H	J			
10.	(913,003)	912,999			

T	H	E	Y
1	2	3	4

L	A	U	N	C	H
5	6	7	8	9	10

H	A	I	R	B	A	L	L	S
11	12	13	14	15	16	17	18	19

10   **Holt Mathematics**

Practice A
Estimating with Whole Numbers

Round each number to the greatest place value.

1. 67 __70__   **2.** 81 __80__   **3.** 24 __20__

4. 115 __100__   **5.** 575 __600__   **6.** 1,852 __2,000__

Estimate each sum or difference. Possible answers:

7. 42 + 19   **8.** 63 − 28   **9.** 37 + 34

____60____   ____30____   ____70____

10. 93 − 14   **11.** 104 + 178   **12.** 112 − 9

____80____   ____300____   ____100____

Estimate each product.

13. 2 × 19   **14.** 87 × 2   **15.** 26 × 3

____40____   ____180____   ____75____

Rewrite each problem using compatible numbers. Then divide.

16. 148 ÷ 5   **17.** 412 ÷ 4   **18.** 70 ÷ 6

__150 ÷ 5; 30__   __400 ÷ 4; 100__   __72 ÷ 6; 12__

19. 62 ÷ 3   **20.** 40 ÷ 7   **21.** 29 ÷ 4

__63 ÷ 3; 21__   __42 ÷ 7; 6__   __28 ÷ 4; 7__

22. A fin whale weighs 44 tons. A gray whale weighs 32 tons. About how much more does a fin whale weigh than a gray whale?

__about 10 tons more__

23. The Suez Canal in Egypt is 108 miles long. The Erie Canal in New York is 363 miles long. About how long are the two canals together?

__about 500 miles long__

11   **Holt Mathematics**

Practice B
Estimating with Whole Numbers

Estimate each sum or difference. Possible answers:

1. 67 + 14   **2.** 583 − 329   **3.** 94 − 36

____80____   ____300____   ____50____

4. 2,856 + 2,207   **5.** 276 + 316   **6.** 6,020 − 3,688

__5,000__   ____600____   __2,000__

7. 34,465 + 19,002   **8.** 78,135 − 19,431   **9.** 216,135 + 165,800

__50,000__   __60,000__   __400,000__

Estimate each product or quotient.

10. 59 ÷ 6   **11.** 51 × 8   **12.** 83 ÷ 4

____10____   ____400____   ____21____

13. 9 × 27   **14.** 49 ÷ 6   **15.** 53 × 8

____270____   ____8____   ____400____

16. 147 ÷ 5   **17.** 118 ÷ 6   **18.** 79 × 5

____30____   ____20____   ____400____

19. Sailfish are the fastest fish in the world. They can swim 68 miles an hour. About how far can a sailfish swim in 3 hours?

__about 210 miles__

20. At a height of 3,281 feet, Angel Falls in Venezuela is the tallest waterfall in the world. Niagara Falls in the United States is only 190 feet tall. About how much taller is Angel Falls?

__about 3,000 feet taller__

21. Ali, a gardener, is preparing to fertilize a lawn. The lawn is 30 yards by 25 yards. One bag of fertilizer will cover an area of 100 square yards. How many bags of fertilizer does Ali need to buy?

__8 bags__

12   **Holt Mathematics**

Practice C
Estimating with Whole Numbers

Estimate each sum or difference. Possible answers:

1. 651 + 124   **2.** 344 − 175   **3.** 1,862 + 1,403

____800____   ____100____   __3,000__

4. 25,661 + 11,706   **5.** 59,210 − 24,337   **6.** 542,901 + 251,504

__40,000__   __40,000__   __800,000__

7. 346,132 − 131,649   **8.** 292,126 + 167,165   **9.** 912,910 − 315,904

__200,000__   __500,000__   __600,000__

Estimate each product or quotient.

10. 76 × 3   **11.** 124 ÷ 3   **12.** 57 × 4

____240____   ____40____   ____240____

13. 538 ÷ 61   **14.** 359 ÷ 64   **15.** 179 × 21

____9____   ____6____   __4,000__

16. 8 × 56   **17.** 263 ÷ 13   **18.** 9 × 63

____480____   ____20____   ____540____

19. The greatest depth of the Sea of Japan is 12,276 feet. The Bering Sea is 3,383 feet deeper than the Sea of Japan. The Caribbean Sea is 7,129 feet deeper than the Bering Sea. About how deep is the Bering Sea? the Caribbean Sea?

__about 15,000 feet; about 22,000 feet__

20. Sperm whales dive up to 7,476 feet in search of food, which is about 9 times deeper than emperor penguins dive. About how deep do the penguins dive?

__about 800 feet__

13   **Holt Mathematics**

Reteach
Estimating with Whole Numbers

In mathematics, you can find an estimate when an exact answer is not needed. An estimate is close to the exact answer.

You can use rounding to estimate sums and differences.

A. Estimate the sum by rounding to the hundreds.

$$\begin{array}{rcl} 3,478 & \longrightarrow & 3,500 \\ + 7,136 & \longrightarrow & + 7,100 \\ \hline & & 10,600 \end{array}$$

B. Estimate the difference by rounding to the thousands.

$$\begin{array}{rcl} 23,848 & \longrightarrow & 24,000 \\ - 16,132 & \longrightarrow & - 16,000 \\ \hline & & 8,000 \end{array}$$

Estimate each sum or difference by rounding to the place value indicated.

1. hundreds

$$\begin{array}{rcl} 789 & \longrightarrow & 800 \\ + 453 & \longrightarrow & + 500 \\ \hline & & 1,300 \end{array}$$

2. thousands

$$\begin{array}{rcl} 4,987 & \longrightarrow & 5,000 \\ - 2,348 & \longrightarrow & - 2,000 \\ \hline & & 3,000 \end{array}$$

3. tens

$$\begin{array}{rcl} 456 & \longrightarrow & 460 \\ + 875 & \longrightarrow & + 880 \\ \hline & & 1,340 \end{array}$$

4. tens

$$\begin{array}{rcl} 876 & \longrightarrow & 880 \\ - 432 & \longrightarrow & - 430 \\ \hline & & 450 \end{array}$$

5. hundreds

$$\begin{array}{rcl} 6,898 & \longrightarrow & 6,900 \\ + 2,671 & \longrightarrow & + 2,700 \\ \hline & & 9,600 \end{array}$$

6. thousands

$$\begin{array}{rcl} 1,857 & \longrightarrow & 2,000 \\ + 3,598 & \longrightarrow & + 4,000 \\ \hline & & 6,000 \end{array}$$

7. hundreds

$$\begin{array}{rcl} 8,813 & \longrightarrow & 8,800 \\ - 2,384 & \longrightarrow & - 2,400 \\ \hline & & 6,400 \end{array}$$

8. thousands

$$\begin{array}{rcl} 9,128 & \longrightarrow & 9,000 \\ - 4,716 & \longrightarrow & - 5,000 \\ \hline & & 4,000 \end{array}$$

14   **Holt Mathematics**

Reteach
Estimating with Whole Numbers (continued)

You can use rounding and basic facts to estimate products. Count the number of zeros in your rounded numbers. They will appear to the right of the basic fact in your estimate.

Estimate 8×532.

8×532
↓   ↓
8×500 Round each factor.
↓   ↓   two zeros
$4,000$

Use rounding to estimate each product.

9. 28×5
10. 78×11
11. 67×19
12. 93×7

| 150 | 800 | 1,400 | 630 |

Compatible numbers are numbers that are easy to compute mentally. One compatible number divides evenly into the other.

Estimate the quotient of $553 \div 8$.

Step 1: What are the multiples of 8?
8 16 24 32 40 48 56 64
Which multiple is closest to 55?
56 is close to 55.
8 and 560 are compatible numbers.

Step 2: Divide. $560 \div 8 = 70$

Use compatible numbers to estimate each quotient.

13. $748 \div 25$ 14. $557 \div 8$ 15. $417 \div 7$ 16. $241 \div 3$

| 30 | 70 | 60 | 80 |

15
Holt Mathematics

Challenge
A Shopping Spree!

You have just won a $2,000 shopping spree at Electronics City! Use estimation and the store's advertisement below to make two different shopping lists of what you can buy without going over your spending limit.

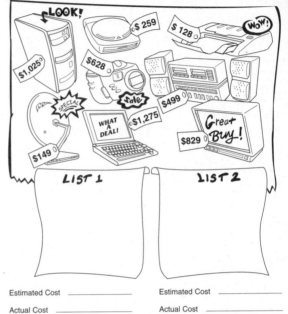

Estimated Cost _____   Estimated Cost _____

Actual Cost _____   Actual Cost _____

Answers will vary depending on students' chosen items. Check for correct estimation. All lists should total less than $2,000.

16
Holt Mathematics

Problem Solving
Estimating with Whole Numbers

Use the table below to answer each question.

Facts About the World's Oceans

Ocean	Area (square mi)	Greatest Depth (ft)
Arctic	5,108,132	18,456
Atlantic	33,424,006	30,246
Indian	28,351,484	24,460
Pacific	64,185,629	35,837

1. If the depths of all the oceans were rounded to the nearest ten thousand, which two oceans would have the same depth?

   Arctic and Indian

2. In 1960, scientists observed sea creatures living as far down as thirty thousand feet. In which ocean(s) could these creatures have lived?

   Pacific and Atlantic

3. If you wanted to compare the depths of the Pacific Ocean and the Atlantic Ocean, which place value would you use to estimate?

   thousands

4. The oceans cover about three-fourths of Earth's surface. Estimate the total area of all the oceans combined by rounding to the nearest million.

   about 130 million sq. mi

Choose the letter for the best answer.

5. There are 5,280 feet in a mile. About how many miles deep is the deepest point in the Pacific Ocean?

   A about 0.7 mile   C about 70 miles
   B about 7 miles   D about 700 miles

6. Rounding to the greatest place value, about how much larger is the Indian Ocean than the Arctic Ocean?

   F about 5 million sq. mi
   G about 10 million sq. mi
   H about 15 million sq. mi
   J about 25 million sq. mi

7. The Atlantic Ocean is about 40 times larger than the world's largest island, Greenland. Use this information to estimate the area of Greenland.

   A about 800,000 sq. mi
   B about 8,000,000 sq. mi
   C about 80,000,000 sq. mi
   D about 1,200,000,000 sq. mi

8. About how much larger would the Pacific Ocean have to be to have more area than the other three oceans combined?

   F about 2 hundred sq. mi
   G about 2 thousand sq. mi
   H about 2 million sq. mi
   J about 20 million sq. mi

17
Holt Mathematics

Reading Strategies
Draw Conclusions

In daily situations that involve math problems, an estimate is sometimes used rather than an exact answer. An **estimate** is an answer that **is close to the exact number.** Read these statements that give estimates:
- Over 45,000 fans attended the opening baseball game.
- The cost of admission is about $10.
- According to the map, we must drive about 50 miles.

In some situations, it is better to **overestimate**. Examples:
- the amount of money to take to the baseball game
- the driving time to the game

In these situations, an overestimate is best. This ensures that you have enough money and arrive at the game on time.

In other situations an **underestimate** would be best. Examples:
- the weight the ballpark express elevator can hold
- the number of "standing room only" tickets available

In these situations, an underestimate is best. This ensures that the elevator is not too heavy and that the "standing room only" section is not too crowded.

Tell whether an overestimate or an underestimate is best for each situation and why.

1. The weight that an airplane can hold.

   underestimate; don't want to overload the plane for safety

2. The amount of money for a trip.

   overestimate; don't want to run out of money

3. The number of people at a track meet.

   overestimate or underestimate; if you wanted people to think that
   the meet was well attended, you would overestimate

4. The number of people who can sit in a section of bleachers.

   underestimate; make sure you have enough seats for fans

5. The number of hours to drive from Chicago to New York.

   overestimate; want to allow for more than enough
   driving time

18
Holt Mathematics

Puzzles, Twisters & Teasers
The Great Race

Sam and Lloyd are bicycle racing around Europe starting in Lisbon, Portugal. The race rules state that Sam and Lloyd must each travel a different route between cities. The final distances between cities for both Sam and Lloyd are listed below.

1. For each city, round the distance for each biker to the nearest 100.

2. Add the distance to the bikers' totals.

3. The race winner is the biker who completes the race with the lowest total distance (rounded to the nearest 100 miles).

4. As an added bonus, at each city an alphabet letter accompanies each racer's total. Put the letter from the winning racer at each city into the blanks below to solve the riddle. For example, Sam traveled a total of 300 miles to Madrid and Lloyd traveled a total of 400 miles. So write the "a" from Sam's total box in the first blank in the riddle.

Origin City	Sam's Route Distance	Lloyd's Route Distance	Sam's Total		Lloyd's Total	
Lisbon, Portugal	—	—	0		0	
Madrid, Spain	345	380	300	a	400	o
Paris, France	742	795	700	p	800	r
London, England	350	140	400	t	100	e
Brussels, Belgium	251	249	300	e	200	a
Berlin, Germany	453	357	500	f	400	p
Vienna, Austria	461	449	500	k	400	p
Athens, Greece	710	854	700	e	900	g
Rome, Italy	538	587	500	a	600	i
Lisbon, Portugal	1,171	1,046	1,200	r	1,000	l

AND THE WINNER IS: _____Lloyd_____ !

Riddle: What do you call a gorilla with a banana?

An a p e with a p p e a l !

19

Practice A
Exponents

Name the base and the exponent for each of the following.

1. 7^2 base __7__ exponent __2__

2. 5^4 base __5__ exponent __4__

3. 6^8 base __6__ exponent __8__

4. 5^9 base __5__ exponent __9__

5. 10^7 base __10__ exponent __7__

6. 4^3 base __4__ exponent __3__

Write using exponents.

7. 4×4
4^2

8. $2 \times 2 \times 2$
2^3

9. 10×10
10^2

10. $5 \times 5 \times 5 \times 5$
5^4

11. $3 \times 3 \times 3 \times 3$
3^4

12. $8 \times 8 \times 8 \times 8 \times 8$
8^5

Write as repeated multiplication.

13. 6^2
6×6

14. 5^3
$5 \times 5 \times 5$

15. 10^3
$10 \times 10 \times 10$

16. 9^4
$9 \times 9 \times 9 \times 9$

17. 2^5
$2 \times 2 \times 2 \times 2 \times 2$

18. 3^6
$3 \times 3 \times 3 \times 3 \times 3 \times 3$

19. How many different ways can you use the digits 3 and 5 to write expressions in exponential form? What are the expressions?
two ways; 3^5 and 5^3

20. What do the following two expressions have in common? "three to the second power" and "three squared"
They both mean 3×3, or 3^2.

20

Practice B
Exponents

Write each expression in exponential form.

1. 9×9
9^2

2. $7 \times 7 \times 7$
7^3

3. $1 \times 1 \times 1 \times 1 \times 1$
1^5

4. $5 \times 5 \times 5 \times 5$
5^4

5. $2 \times 2 \times 2 \times 2 \times 2 \times 2$
2^6

6. $10 \times 10 \times 10 \times 10$
10^4

Find each value.

7. 6^2
36

8. 5^3
125

9. 10^3
1,000

10. 7^2
49

11. 2^5
32

12. 3^4
81

13. 25^1
25

14. 16^0
1

Compare. Write <, >, or =.

15. 8^0 < 7^1

16. 10^2 < 11^2

17. 8^2 = 4^3

18. 3^4 > 5^2

19. 2^5 < 9^2

20. 6^2 > 3^3

21. What whole number equals 25 when it is squared and 125 when it is cubed?
5

22. Use exponents to write the number 81 three different ways.
81^1; 9^2; 3^4

21

Practice C
Exponents

Write each expression in exponential form.

1. $10 \times 10 \times 10 \times 10$
10^4

2. $7 \times 7 \times 7 \times 7 \times 7$
7^5

3. $4 \times 4 \times 4$
4^3

Find each value.

4. 8^2 __64__

5. 4^3 __64__

6. 6^3 __216__

7. 15^2 __225__

8. 2^8 __256__

9. 3^5 __243__

10. 38^1 __38__

11. 7^3 __343__

Compare. Write <, >, or =.

12. 8^2 = 4^3

13. 9^2 > 5^2

14. 6^2 < 3^4

15. 7^2 > 2^4

16. 10^2 = 100^1

17. 81^0 < 9^2

18. $4^2 + 5$ > $3^3 - 7$

19. $2^3 + 2$ > $3^2 - 2$

20. $2^5 - 10$ = $4^2 + 6$

21. If it takes Cell A 3 hours to produce two cells, how many cells will Cell A produce in 24 hours?
256 cells

22. Use exponents to complete the table.

Generation	Number of People	Exponent
Parents	2	2^1
Grandparents	4	2^2
Great Grandparents	8	2^3
Great-Great Grandparents	16	2^4
Great-Great-Great Grandparents	32	2^5

22

Reteach
Exponents

You can write a number in exponential form to show repeated multiplication. A number written in exponential form has a base and an exponent. An exponent tells you how many times a number, called the base, is used as a factor.

8^4 ← exponent

base

Write the expression in exponential form.
$6 \times 6 \times 6$
6 is used as a factor 3 times.
$6 \times 6 \times 6 = 6^3$

Write each expression in exponential form.

1. $8 \times 8 \times 8 \times 8 \times 8$ 2. 3×3 3. $5 \times 5 \times 5 \times 5$ 4. $7 \times 7 \times 7$

8^5 3^2 5^4 7^3

You can find the value of expressions in exponential form. Find the value.
2^5

Step 1: Write the expression as repeated multiplication.
$2^5 = 2 \times 2 \times 2 \times 2 \times 2$

Step 2: Multiply.
$2 \times 2 \times 2 \times 2 \times 2 = 32$

$2^5 = 32$

Find each value.

5. 12^3 6. 6^5 7. 10^4 8. 4^6

__1,728__   __7,776__   __10,000__   __4,096__

Holt Mathematics

Challenge
Exponent Riddle

What is the greatest number that can be written with two digits?

Find the value of each expression below. Then in the box at the bottom of the page, write each expression's letter in the blank above its value. When you have found all the values, you will have solved the riddle.

E 3^3 ____27____

H 5^2 ____25____

I 2^4 ____16____

N 34^0 ____1____

O 9^2 ____81____

P 4^3 ____64____

R 6^2 ____36____

T 7^2 ____49____

W 10^2 ____100____

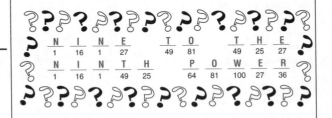

N I N E   T O   T H E
1 16 1 27   49 25   27

N I N T H   P O W E R
1 16 1 49 25   64 81 100 27 36

Holt Mathematics

Problem Solving
Exponents

1. The Sun is the center of our solar system. The Sun is the star closest to our planet. The surface temperature of the Sun is close to 10,000°F. Write 10,000 using exponents.

__10^4__

2. Patty Berg has won 4^2 major women's titles in golf. Write 4^2 in standard form.

__16__

3. William has 3^3 baseball cards and 4^3 football cards. Write the number of baseball cards and footballs cards that William has.

__27 baseball cards and__

__64 football cards__

4. Michelle recorded the number of miles she ran each day last year. She used the following expression to represent the total number of miles: $3 \times 3 \times 3 \times 3 \times 3 \times 3 \times 3$. Write this expression using exponents. How many miles did Michelle run last year?

__3^7; 2,187 miles__

Choose the letter for the best answer.

5. In Tyrone's science class he is studying cells. Cell A divides every 30 minutes. If Tyrone starts with two cells, how many cells will he have in 3 hours?
   A 6 cells
   B 32 cells
   Ⓒ 128 cells
   D 512 cells

6. Tanisha's soccer team has a phone tree in case a soccer game is postponed or cancelled. The coach calls 2 families. Then each family calls 2 other families. How many families will be notified during the 4th round of calls?
   F 2 families
   G 4 families
   H 8 families
   Ⓙ 16 families

7. The Akashi-Kaiko Bridge is the longest suspension bridge in the world. It is located in Kobe-Naruto, Japan and was completed in 1998. It is about 3^8 feet long. Write the approximate length of the Akashi-Kaiko Bridge in standard form.
   Ⓐ 6,561 feet
   B 2,187 feet
   C 512 feet
   D 24 feet

8. The Strahov Stadium is the largest sports stadium in the world. It is located in Prague, Czech Republic. Its capacity is about 12^5 people. Write the capacity of the Strahov Stadium in standard form.
   F 60 people
   G 144 people
   H 20,736 people
   Ⓙ 248,832 people

Holt Mathematics

Reading Strategies
Synthesize Information

Exponents are an efficient way to write repeated multiplication.

Read 2^4 → *2 to the fourth power*

   2^4 means **2 is a factor 4 times,** or $2 \times 2 \times 2 \times 2$

Read $2^4 = 16$ → *2 to the fourth power equals 16.*

Exponent	Meaning	Value
10^3 *10 to the third power*	10 is a factor 3 times: $10 \times 10 \times 10$	$10^3 = 1,000$
6^5 *6 to the fifth power*	6 is a factor 5 times: $6 \times 6 \times 6 \times 6 \times 6$	$6^5 = 7,776$

Answer each question.

1. Write in words how you would read 3^4. __three to the fourth power__

2. What does 3^4 mean? __three is a factor 4 times: $3 \times 3 \times 3 \times 3$__

3. What is the value of 3^4? __$3^4 = 81$__

4. Write in words how you would read 5^3. __five to the third power__

5. Write 5^3 as repeated multiplication. __$5 \times 5 \times 5$__

6. What is the value of 5^3? __$5^3 = 125$__

7. Tell why 2^3 is not 2 x 3.
   __2^3 means $2 \times 2 \times 2$, and 2×3 means $3 + 3$.__

8. Is 3^4 the same as 4^3? __no__ Explain why or why not.
   __Possible answer: $3 \times 3 \times 3 \times 3$ is not the same as $4 \times 4 \times 4$.__

Holt Mathematics

Holt Mathematics

Puzzles, Twisters & Teasers

Answer This!

What are the only land mammals that cannot jump?

To find the answer:

1. Use a ruler to match each number and its value.
   (Each line you draw will cross a number and a letter)

2. Write the letter under the matching number in the decoder.

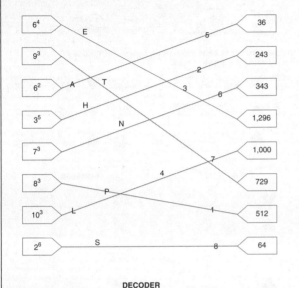

6^4	E		36
9^3		5	243
6^2	A T	2	343
3^5	H	3 6	1,296
7^3	N		1,000
8^3		7	729
10^3	P	4	512
2^6	L		64
	S	8	

DECODER

3	4	3	1	2	5	6	7	8
E	L	E	P	H	A	N	T	S

27   **Holt Mathematics**

Practice A

Order of Operations

Name the operation you should perform first.

1. $5 + 6 \times 2$ **2.** $18 \div 3 - 1$ **3.** $4 + (7 - 1)$

_____ multiply _____   _____ divide _____   _____ subtract _____

4. $3^2 + 6$ **5.** $(15 + 38) \times 6$ **6.** $5 \times 10 - 12$

_____ exponent _____   _____ add _____   _____ multiply _____

Match each expression to its value.

	Expression	Value
G	**7.** $7 + 8 - 2$	**A.** 9
F	**8.** $9 + (12 - 10)$	**B.** 40
H	**9.** $(20 - 15) \times 2$	**C.** 12
A	**10.** $10 \div 5 + 7$	**D.** 0
C	**11.** $6 + 2 \times 3$	**E.** 16
E	**12.** $(2 \times 4) + 8$	**F.** 11
D	**13.** $14 \div 2 \times 0$	**G.** 13
B	**14.** $(5 - 1) \times 10$	**H.** 10

15. Sam bought two CDs for $13 each. Sales tax for both CDs was
$3. Write an expression to show how much Sam paid in all.

$(2 \times 13) + 3$

16. Alicia made 24 chocolate chip cookies and 36 sugar cookies.
Then she divided all the cookies into 10 bags to sell at the
bake sale. Write an expression to show how many cookies she put
into each bag.

$(24 + 36) \div 10$

28   **Holt Mathematics**

Practice B

Order of Operations

Evaluate each expression.

1. $10 + 6 \times 2$ **2.** $(15 + 39) \div 6$ **3.** $(20 - 15) \times 2 + 1$

_____ 22 _____   _____ 9 _____   _____ 11 _____

4. $(4^2 + 6) \div 11$ **5.** $9 + (7 - 1) \times 2$ **6.** $(2 \times 4) + 8 - (5 \times 3)$

_____ 2 _____   _____ 21 _____   _____ 1 _____

7. $5 + 18 \div 3^2 - 1$ **8.** $8 + 5 \times 10 - 12$ **9.** $14 + (50 - 7^2) \times 3$

_____ 6 _____   _____ 46 _____   _____ 17 _____

Add parentheses so that each equation is correct.

10. $7 + 9 \times 3 - 1 = 25$ **11.** $2^3 - 7 \times 4 = 4$ **12.** $5 + 6 \times 9 \div 3 = 23$

_____ $(3 - 1)$ _____   _____ $(2^3 - 7)$ _____   _____ $(9 \div 3)$ _____

13. $12 \div 3 \times 2 = 2$ **14.** $8 + 3 \times 6 - 4 - 1 = 13$ **15.** $4 \times 3^2 + 1 = 40$

_____ $(3 \cdot 2)$ _____   _____ $(6 - 4)$ _____   _____ $(3^2 + 1)$ _____

16. $9 \times 0 + 5 - 3 = 42$ **17.** $15 \times 3^2 - 2^3 = 15$ **18.** $14 \div 2 + 5 \times 5 = 10$

_____ $(0 + 5)$ _____   _____ $(3^2 - 2^3)$ _____   _____ $(2 + 5)$ _____

19. Tyler walked 2 miles a day for the first week of his exercise plan.
Then he walked 3 miles a day for the next 9 days. How many
miles did Tyler walk in all?

41 miles

20. Paulo's father bought 8 pizzas and 12 bottles of juice for the
class party. Each pizza cost $9 and each bottle of juice cost $2.
Paulo's father paid with a $100-bill. How much change did he
get back?

$4

29   **Holt Mathematics**

Practice C

Order of Operations

Evaluate each expression.

1. $42 - 3 \times 10 + 2$ **2.** $1 + 4^3 - 16$ **3.** $(15 - 6) \times 2 + 20$

_____ 14 _____   _____ 49 _____   _____ 38 _____

4. $(5^2 + 3^2 + 2) \div 6$ **5.** $61 - 5 \times 2^3 + 5$ **6.** $7 \times 8 + (2 \times 4) \div 2^2$

_____ 6 _____   _____ 26 _____   _____ 58 _____

Add parentheses so that each equation is correct.

7. $12 - 3 \times 2 + 4^2 = 34$ **8.** $72 \div 2 \times 4 \div 3 = 3$

_____ $(12 - 3)$ _____   _____ (2×4) _____

9. $13 + 7 - 6 + 4 \times 2 = 0$ **10.** $28 \div 7 + 3^3 - 3^2 - 1 = 21$

_____ $(6 + 4)$ _____   _____ $(3^3 - 3^2)$ _____

**Use each of the numbers 2, 3, 4, and 6 once to make each
equation correct. Possible answers are given.**

11. $(\_\_ - \_\_) + \_\_ \times \_\_ = 11$ **12.** $\_\_ \times \_\_ - (\_\_ \div \_\_) = 6$

_____ $(6 - 3) + 2 \times 4$ _____   _____ $2 \times 4 - (6 \div 3)$ _____

13. $\_\_ + (\_\_ \times \_\_) \times \_\_ = 30$ **14.** $\_\_ \div \_\_ + \_\_ \times \_\_ = 20$

_____ $6 + (2 \times 3) \times 4$ _____   _____ $4 \div 2 + 6 \times 3$ _____

15. Use an exponent to write an expression with five 3s that has a
value of 0.

Possible answer: $3 \times 3 \times 3 - 3^3$

16. Mrs. Thompson is putting new tile on her bathroom floor. Each
tile measures 2 inches on each side. The bathroom floor is
3 feet long and 2 feet wide. How many tiles will she use to
cover the entire floor?

216 tiles

30   **Holt Mathematics**

Reteach
Order of Operations

A mathematical phrase that includes only numbers and operations is called a numerical expression.

$9 + 8 \times 3 \div 6$ is a numerical expression.

To evaluate a numerical expression, you find its value.

You can use the order of operations to evaluate a numerical expression.

Order of Operations
1. Do all operations within parentheses.
2. Find the values of numbers with exponents.
3. Multiply and divide in order from left to right.
4. Add and subtract in order from left to right.

Evaluate the expression.

$60 \div (7 + 3) + 3^2$	
$60 \div 10 + 3^2$	Do all operations within parentheses.
$60 \div 10 + 9$	Find the values of numbers with exponents.
$6 + 9$	Multiply and divide in order from left to right.
15	Add and subtract in order from left to right.

Evaluate each expression.

1. $7 \times (12 + 8) - 6$

$7 \times \underline{20} - 6$

$\underline{140} - 6$

$\underline{\hspace{1cm}134\hspace{1cm}}$

2. $10 \times (12 + 34) + 3$

$10 \times \underline{46} + 3$

$\underline{460} + 3$

$\underline{\hspace{1cm}463\hspace{1cm}}$

3. $10 + (6 \times 5) - 7$

$10 + \underline{30} - 7$

$\underline{40} - 7$

$\underline{\hspace{1cm}33\hspace{1cm}}$

4. $2^3 + (10 - 4)$

$\underline{\hspace{1cm}14\hspace{1cm}}$

5. $7 + 3 \times (8 + 5)$

$\underline{\hspace{1cm}46\hspace{1cm}}$

6. $36 \div 4 + 11 \times 8$

$\underline{\hspace{1cm}97\hspace{1cm}}$

7. $5^2 - (2 \times 8) + 9$

$\underline{\hspace{1cm}18\hspace{1cm}}$

8. $3 \times (12 \div 4) - 2^2$

$\underline{\hspace{1cm}5\hspace{1cm}}$

9. $(3^3 + 10) - 2$

$\underline{\hspace{1cm}35\hspace{1cm}}$

Holt Mathematics

Challenge
Crack the Expression Code

Each of these symbols stands for a different operation symbol:

Each of these animals stands for a different whole number 1–4:

Use the equations below to find what each symbol and animal represents in the expression code.

 = **7**

OPERATIONS

1. ♥ = _____+_____

2. ♠ = _____−_____

3. ♦ = _____×_____

4. ♣ = _____÷_____

NUMBERS

5. = _____4_____

6. = _____2_____

7. = _____1_____

8. = _____3_____

Holt Mathematics

Problem Solving
Order of Operations

Evaluate each expression to complete the table.

Mammals with the Longest Tails

	Mammal	Expression	Tail Length
1.	Asian elephant	$2 + 3^2 \times 7 - (10 - 4)$	59
2.	Leopard	$5 \times 6 + 5^2$	55
3.	African elephant	$6 \times (72 \div 8) - 3$	51
4.	African buffalo	$51 + 6^2 \div 9 - 12$	43
5.	Giraffe	$4^3 - 3 \times 7$	43
6.	Red kangaroo	$11 + 48 \div 6 \times 4$	43

Choose the letter for the best answer.

7. Adam and his two brothers went to the zoo. Each ticket to enter the zoo costs $7. Adam bought two bags of peanuts for $4 each, and one of his brothers bought a lion poster for $12. Which expression shows how much money they spent at the zoo in all?

A $7 + 4 + 12$

B $7 \times 3 + 4 + 12$

C $7 \times 3 + 4 \times 2 + 12$

D $(7 \times 3) + (4 \times 12)$

8. An elephant eats about 500 pounds of grass and leaves every day. There are 2 Africa elephants and 3 Asian elephants living in the City Zoo. How many pounds of grass and leaves do the zookeepers need to order each week to feed all the elephants?

F 2,500 pounds

G 17,500 pounds

H 3,000 pounds

J 21,000 pounds

9. The average giraffe is 18 feet tall. Which of these expressions shows the height of a giraffe?

A $4^2 - 2$

B $3 \times 12 \div 4 + 2$

C $3^3 \div 9 \times 6$

D $20 \div 5 + 5 - 6$

10. Some kangaroos can cover 30 feet in a single jump! If a kangaroo could jump like that 150 times in a row, how much farther would it need to go to cover a mile? (1 mile = 5,280 feet)

F 780 feet

G 26 feet

H 176 feet

J 5,100 feet

Holt Mathematics

Reading Strategies
Use a Flowchart

When you read a sentence, you read each word in order from left to right. To evaluate an expression, you cannot always compute the operations in the order they are given, from left to right. You must follow the order of operations. The order is listed in the flowchart below:

Parentheses → Exponents → Multiply or divide from left to right → Add or subtract from left to right

Example $\quad 3 \times (5 + 4) - 2^2$ **Parentheses**

$\qquad 3 \times 9 - 2^2$ **Exponents**

$\qquad 3 \times 9 - 4$ **Multiply or divide (left to right)**

$\qquad 27 - 4$ **Add or subtract (left to right)**

$\qquad 23$

Answer each question.

1. Write a flowchart to list the order in which you would compute the operations for this expression: $12 - 2 \times 3 + 8 \div 2$.

Multiply → Divide → Subtract → Add

2. Evaluate this expression: $12 - 2 \times 3 + 8 \div 2$.

10

3. Write a flowchart to list the order in which you would compute the operations for this expression: $(12 - 2) \times 3 + 8 \div 2$.

Parentheses → Multiply → Divide → Add

4. Evaluate this expression: $(12 - 2) \times 3 + 8 \div 2$.

34

5. Explain how Exercises 1 and 3 are alike and how they are different.

Possible explanation: The numbers and operations in each expression are the same, but the value of each expression is different since Exercise 3 includes parentheses.

Holt Mathematics

Holt Mathematics

Puzzles, Twisters & Teasers
Are You in Order?

What did one telephone say to the other when it proposed?

To answer the riddle, solve the following problems. Then write
the letter that is represented by each answer in the blanks below.

1. $20 + 16 \times 2 =$ ___52___ (I)

2. $55 \div (11 - 6) \times 8 =$ ___88___ (W)

3. $4 + 9 - (2 + 6) + 3 =$ ___8___ (L)

4. $(24 + 12) \div 12 =$ ___3___ (G)

5. $(4 + 6 \div 2) \times (1 + 9) =$ ___70___ (V)

6. $10 \times (54 - 49) + 17 =$ ___67___ (E)

7. $(36 \div 18)^3 + 17 \times 3 =$ ___59___ (Y)

8. $2^4 + (81 - 50) + 52 =$ ___99___ (O)

9. $21 \div (2 + 1) \times 5 - 2^2 =$ ___31___ (U)

10. $6 \div (1 + 2) \times 5^2 - 25 =$ ___25___ (A)

11. $32 \times (3 + 2) + 8 \div 2 =$ ___164___ (R)

12. $(6^3 \div 3) + 8 \div 2 =$ ___76___ (N)

I		W	I	L	L
88		52	8	8	

G	I	V	E		Y	O	U
3	52	70	67		59	99	31

A		R	I	N	G
25		164	52	76	3

35   **Holt Mathematics**

Practice A
Mental Math

Choose the letter of the equation that shows the given property.

1. Associative Property
A $2 + 3 = 3 + 2$
B $7 \times 8 = 7 \times (4 + 4)$
C $8 \times (6 \times 5) = (8 \times 6) \times 5$ *(circled)*
D $9 \times (2 + 4) = (9 \times 2) + (9 \times 4)$

2. Distributive Property
F $3 \times (6 \times 11) = (3 \times 6) \times 11$
G $75 + 15 = 15 + 75$
H $9 \times 8 = 8 \times 9$
J $12 \times (4 + 7) = (12 \times 4) + (12 \times 7)$ *(circled)*

3. Commutative Property
A $3 \times (7 + 8) = 3 \times 15$
B $(10 + 4) + 3 = 10 + (4 + 3)$
C $(9 + 2) \times 5 = (9 \times 5) + (2 \times 5)$
D $6 \times 5 = 5 \times 6$ *(circled)*

4. Associative Property
F $20 \times (3 + 3) = (20 \times 3) + (20 \times 3)$
G $4 + (3 + 9) = (4 + 3) + 9$ *(circled)*
H $(10 + 5) \times 7 = 15 \times 7$
J $16 \times 8 = 8 \times 16$

Rewrite each expression using the named property.

5. $8 + 12$; Commutative Property
$12 + 8$

6. $(9 \times 6) \times 4$; Associative Property
$9 \times (6 \times 4)$

7. $3 \times (5 + 2)$; Distributive Property
$(3 \times 5) + (3 \times 2)$

8. $2 \times (4 + 5)$; Distributive Property
$(2 \times 4) + (2 \times 5)$

Find each sum or product.

9. $7 + 15 + 3 + 5$ 30

10. $7 \times 2 \times 5$ 70

11. $4 \times 3 \times 5$ 60

Multiply using the Distributive Property.

12. 4×38 152

13. 6×53 318

14. 8×42 336

15. Sue has \$4, Tom has \$11, Brian has \$6, and Anita has \$9. Use
mental math to find how much money they have altogether.
\$30

16. Each minibus seats 14 people, and the school owns 5 minibuses.
Use mental math to find how many students can ride in the
school's minibuses at the same time.
70 students

36   **Holt Mathematics**

Practice B
Mental Math

Evaluate.

1. $17 + 4 \times 5$ 37

2. $25 \times 3 \times 4$ 300

3. $28 + 39 + 11 + 22$ 100

4. $12 + 7 + 8 + 13$ 40

5. $10 + 3 \times 2$ 16

6. $9 \times 8 \times 5$ 360

7. $97 + 4 + 3 + 26$ 130

8. $2 \times 6 \times 5$ 60

9. $28 + 2 \times 6$ 40

Use the Distributive Property to find each product.

10. 4×16 64

11. 8×31 248

12. 3×62 186

13. 2×46 92

14. 5×29 145

15. 7×22 154

16. 9×21 189

17. 6×15 90

18. 8×44 352

19. 4×29 116

20. 7×31 217

21. 5×57 285

22. Each ticket to a play costs \$27. How much will it cost to buy
4 tickets? Which property did you use to solve this problem
with mental math?
\$108; Distributive Property

23. Mr. Stanley bought two cases of pencils. Each case has 20 boxes.
In each box there is 10 pencils. Use mental math to find how
many pencils Mr. Stanley bought.
400 pencils

24. When you consider that cows eat grass and the water needed to
grow the grass that cows eat, it takes 65 gallons of water to
produce one serving of milk! Use mental math to find how many
gallons of water are needed to produce 5 servings of milk.
325 gallons

37   **Holt Mathematics**

Practice C
Mental Math

Let a, b, and c represent three different numbers. Use them to
write equations showing each property.

1. Associative Property
$(a \times b) \times c = a \times (b \times c)$

2. Commutative Property
$a + b = b + a; a \times b = b \times a$

3. Distributive Property
$a \times (b + c) = (a \times b) + (a \times c)$

Use mental math to find each sum or product.

4. $59 + 27 + 21 + 43$ 150

5. $4 \times 5 \times 8 \times 5$ 800

6. $25 \times 6 \times 5 \times 4$ 3,000

7. $175 + 318 + 82 + 25$ 600

8. $163 + 55 + 37$ 255

9. $5 \times 23 \times 6$ 690

Use the Distributive Property to find each product.

10. 19×7 133

11. 52×40 2,080

12. 62×5 310

13. 32×9 288

14. 11×15 165

15. 7×37 259

16. 4×108 432

17. 84×20 1,680

18. The Cineplex has 13 theaters. Four of the theaters seat
150 people, 8 of the theaters seat 250 people, and the largest
theater seats 275 people. Use mental math to find how many
people can see a movie at the Cineplex at one time.
2,875 people

19. Antoine has two part-time jobs. He earns \$17 an hour working
construction during the day and \$9 an hour stocking shelves in
a hardware store at night. Last week he worked 20 hours at his
construction job and 15 hours at the hardware store. Use mental
math to find how much Antoine earned last week in all.
\$475

38   **Holt Mathematics**

Reteach
Mental Math

Commutative Property
Changing the order of addends does not change the sum.
21 + 13 = 13 + 21

Changing the order of factors does not change the product.
5 × 7 = 7 × 5

Associative Property
Changing the grouping of addends does not change the sum.
(3 + 8) + 4 = 3 + (8 + 4)

Changing the grouping of factors does not change the product.
2 × (7 × 4) = (2 × 7) × 4

Distributive Property
When you multiply a number by a sum, you can
• Find the sum and then multiply.   3 × (8 + 4) = 3 × 12 = 36
                                    or
• Multiply the number by each addend and then find the sum.
  3 × (8 + 4) = (3 × 8) + (3 × 4) = 24 + 12 = 36

Identify the property shown.

1. 3 × (2 × 6) = (3 × 2) × 6
associative

2. 7 + 18 = 18 + 7
commutative

3. 4 × (8 + 5) = 4 × 13
distributive

4. 11 × 8 = 8 × 11
commutative

5. 3 × (8 + 4) = (3 × 8) + (3 × 4)
distributive

6. (3 + 8) + 4 = 3 + (8 + 4)
associative

Identify the property shown and the missing number in each equation.

7. 9 + 16 = y + 9
commutative; y = 16

8. 4 × (3 × 2) = (4 × n) × 2
associative; n = 3

9. 3 × (11 + 4) = 3 × a
distributive; a = 15

10. 6 × (9 + 14) = b × 23
distributive; b = 6

39   Holt Mathematics

Reteach
Mental Math (continued)

Find each sum or product.
A. 8 + 9 + 22 + 31
    8 + 22 + 9 + 31   Use the Commutative Property.
    (8 + 22) + (9 + 31) Use the Associative Property.
       30   +   40   Use mental math to add.
          70

B. 5 × 7 × 4
    7 × 5 × 4   Use the Commutative Property.
    7 × (5 × 4)   Use the Associative Property.
    7 × 20   Use mental math to multiply.
    140

Find each sum or product.

11. 3 + 58 + 27 + 22
110

12. 8 × 3 × 5
120

13. 5 × 3 × 4
60

14. 54 + 32 + 78 + 106
270

15. 84 + 11 + 26 + 39
160

16. 10 × 3 × 7
210

Find the product.
6 × 34

Step 1: Write one factor as a sum of two numbers.
    6 × 34 = 6 × (30 + 4)

Step 2: Use the Distributive Property.
    6 × (30 + 4) = (6 × 30) + (6 × 4)

Step 3: Use mental math to multiply and add.
    (6 × 30) + (6 × 4) = 180 + 24 = 204

Use the Distributive Property to find each product.

17. 6 × 43
258

18. 12 × 34
408

19. 53 × 4
212

20. 74 × 8
592

40   Holt Mathematics

Challenge
Magic Squares

When you add the numbers in each row, each
column, and each diagonal of a magic square,
you get the same number—the magic sum.

In the magic square at the right, for example,
the magic sum is 30.

**Use mental math to complete each magic
square and find the magic sum.**

1.

7	1	10
9	6	3
2	11	5

Magic sum: 18

2.

11	4	9
6	8	10
7	12	5

Magic sum: 24

3.

1	15	14	4
12	6	7	9
8	10	11	5
13	3	2	16

Magic sum: 34

4. Use mental math to create your own
magic square using the numbers 1–9.

2	7	6
9	5	1
4	3	8

Magic sum: Possible answer: 15

41   Holt Mathematics

Problem Solving
Mental Math

The bar graph below shows the average amounts of water used
during some daily activities. Use the bar graph and mental
math to answer the questions.

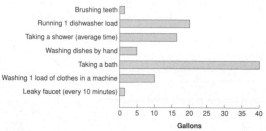

How Much Water?

1. Most people brush their teeth three
times a day. How much water do
they use for this activity every week?
42 gallons

2. How much water is wasted in a day
by a leaky faucet?
288 gallons

3. The average American uses 124 gallons of water a day. Name
a combination of activities listed in the table that would equal
that daily total.
Possible answer: taking a bath, washing 4 loads of laundry,
brushing teeth two times, washing 2 dishwasher loads

Choose the letter for the best answer.

4. Kenya used 24 gallons of water doing three of the activities listed in the table once.
Which activities did she do?

A taking a bath, brushing teeth, washing dishes by hand

B taking a bath, brushing teeth, running 1 dishwasher load

Ⓒ taking a shower, brushing teeth, washing dishes by hand

D taking a shower, brushing teeth, running 1 dishwasher load

5. If you wash two loads of dishes by hand instead of using a dishwasher, how much
water do you save?

Ⓕ 30 gallons   G 15 gallons   H 10 gallons   J 1 gallon

42   Holt Mathematics

Reading Strategies
LESSON 1-5 Focus on Vocabulary

The Commutative, Associative, and Distributive Properties of mathematics can make it easier to use mental math.

Commutative Property—The word **commute** means **to exchange**. In mathematics, when **addends or factors exchange places, the** sum or product is not affected.

Addends change places	Factors change places
13 + 18 + 17	4 × 7 × 5
13 + 17 + 18	4 × 5 × 7
30 + 18 = 48	20 × 7 = 140

Associative Property—The word **associate** means **to join**. In mathematics, **when addends or factors are joined, or grouped, with parentheses** in different ways, the sum or product is not affected.

Addends are grouped	Factors are grouped
11 + 4 + 16	7 × 8 × 5
11 + (4 + 16)	7 × (8 × 5)
11 + 20 = 31	7 × 40 = 280

Distributive Property—The word **distribute** means **to give out**. In mathematics, you can **distribute a factor** over a sum without affecting the original product.

5 × 17	17 = 10 + 7
(5 × 10) + (5 × 7)	Distribute 5 as a factor.
50 + 35	Multiply.
85	Add.

Answer each question.

1. Rewrite 17 + 8 + 13 using the Commutative Property, then compute.

   17 + 13 + 8 = 38

2. Rewrite 9 × 8 × 5 using the Associative Property, then compute.

   9 × (8 × 5) = 360

3. Rewrite 7 × 28 using the Distributive Property, then compute.

   7 × (20 + 8) = 140 + 56 = 196

43   **Holt Mathematics**

Puzzles, Twisters & Teasers
LESSON 1-5 Who Is the Famous Movie Star?

Use the Associative Property to find each sum or product.
Remember: 6 + 13 + 7 = 6 + (13 + 7)   18 × 5 × 2 = 18 × (5 × 2)
          = 6 + 20 = 26   = 18 × 10 = 180

1. 7 + 6 + 4 = __17__   2. 18 + 5 + 5 = __28__

3. 17 + 3 + 10 = __30__   4. 14 + 6 + 12 = __32__

5. 4 × 6 × 10 = __240__   6. 23 + 19 + 11 = __53__

7. 39 × 5 × 2 = __390__   8. 40 × 5 × 3 = __600__

Use the Distributive Property to find the product.
Remember: 3 × 16 = 3 × (10 + 6)
          = (3 × 10) + (3 × 6)
          = 30 + 18 = 48

9. 7 × 15 =   10. 35 × 4 = __140__
   (7 × 10) + (7 × 5)
   __70 + 35 = 105__

11. 56 × 4 = __224__   12. 43 × 7 = __301__

13. 52 × 9 = __468__   14. 71 × 11 = __781__

15. 98 × 12 = __1,176__   16. 222 × 9 = __1,998__

Use the answers to connect the dots to see a famous movie star. Connect the dots in order from the least number to the greatest number. Then connect the dot for the greatest number to the dot for the least number.

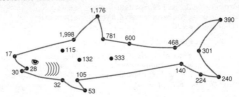

44   **Holt Mathematics**

LESSON Practice A
1-6 Choose the Method of Computation

Answer the questions to describe the method of computation you should use to solve each problem.

PROBLEM: In the 2002 Winter Olympic Games, the United States won 10 gold medals, 13 silver medals, and 11 bronze medals. How many medals did the United States win in all?

1. What method of computation will you use to solve this problem?

   mental math

2. Why did you choose this method of computation?

   Possible answer: There are only 3 numbers to add. The numbers
   are small and one is a multiple of 10.

3. What is the solution to the problem?

   The United States won 34 medals in all.

PROBLEM: The United States holds the record for the most Summer Olympic Medals ever won. As of 2004, the United States had won 850 gold medals, 661 silver medals, and 563 bronze medals. How many Summer Olympic medals has the United States won in all?

4. What method of computation will you use to solve this problem?

   paper and pencil or calculator

5. Why did you choose this method of computation?

   Possible answer: The numbers are too large to add mentally,
   and none of the numbers are compatible.

6. What is the solution to the problem?

   The United States has won 2,074 Summer Olympic medals.

45   **Holt Mathematics**

LESSON Practice B
1-6 Choose the Method of Computation

1. Athletes from 197 countries competed at the 1996 Summer Olympic Games held in Atlanta, Georgia. That is 25 more countries that competed at the 1992 games held in Barcelona, Spain. How many different countries competed in Barcelona?

   Athletes from 172 countries competed in Barcelona.

2. At the 1996 Summer Olympic Games held in Atlanta, Georgia, 10,310 athletes competed. At the 1992 Summer Olympic Games held in Barcelona, Spain, 9,364 athletes competed. How many more athletes competed in Atlanta than in Barcelona?

   946 more athletes competed in Atlanta.

3. The marathon race is one of the oldest events in the Summer Olympic Games. Marathon competitors run a total of 26 miles 385 yards. There are 5,280 feet in a mile and 3 feet in a yard. How many yards long is the entire marathon race?

   The marathon is 46,145 yards long.

4. The world record for the fastest men's marathon race is 2 hours, 5 minutes, 42 seconds. The world record for the fastest women's marathon race is 2 hours, 20 minutes, 43 seconds. How much faster is the men's record marathon time?

   It is 15 minutes, 1 second faster.

5. The men's outdoor world record in the high jump is 2.45 meters or 8 feet 0.5 inches. The women's outdoor world record in the high jump is 2.09 meters or 6 feet 10.25 inches. How much higher is the men's high jump record? Write the answer in meters and feet.

   0.36 meter or 1 foot 2.25 inches

6. The men's world record in the 400-meter relay is 37.40 seconds, held by the U.S. If each of the four runners each ran 100 meters in the same time, how long did each runner run?

   9.35 seconds

7. Athletes from 13 nations competed in the first modern Olympics in 1896. Today, athletes from nearly 200 nations compete in the Summer Olympics. About how many more nations participate in the Olympics today than in 1896?

   about 187 nations

46   **Holt Mathematics**

Practice C
1-6 *Choose the Method of Computation*

Use the information below to answer the questions.

The Paralympics are Olympics Games held for the disabled. The Paralympics are held by the Olympic host country in the same year and usually the same city. The XII Paralympic Summer Games were held in Athens, Greece in 2004. The countries with the most medals are as follows; China 141 (63 gold), Australia 100 (26 gold), and Great Britain 94 (35 gold). Almost 4,000 athletes from 136 nations competed in 19 sports.

1. How many gold medals did the top three countries with the most medals win?

<u>124 gold medals</u>

2. During the 2004 Summer Olympic Games, about 11,000 athletes competed in 28 sports. About how many more athletes competed in the 2004 Summer Olympic Games than in the 2004 Summer Paralympic Games?

<u>about 7,000 more athletes</u>

3. How many more medals did China win than Australia? gold medals?

<u>41 medals; 37 gold medals</u>

4. During the 2004 Summer Olympic Games, the U.S. won the most medals with 103 medals (35 gold). Compare the country with the most medals in the Summer Olympic Games with the country with the most medals in the Summer Paralympic Games. Which country had more medals? how many more? how many more gold medals?

<u>China; 38 medals; 28 more gold medals</u>

5. During the 2004 Summer Olympic Games, China won 63 medals (16 gold). How many medals in all did China take home in 2004? gold medals?

<u>204 medals; 79 gold medals</u>

6. During the 2004 Summer Olympic Games, Great Britain won 30 medals (9 gold). How many more medals in all did Great Britain win in the 2004 Summer Paralympic Games than in the Summer Olympic Games? gold medals?

<u>64 medals; 26 gold medals</u>

47   **Holt Mathematics**

Reteach
1-6 *Choose the Method of Computation*

Paper and pencil, mental math, and a calculator are three computation methods for solving problems.

• If there are many small numbers, use paper and pencil.
• If the numbers are small and easy, use mental math.
• If the numbers are large, use a calculator.

Before you solve a problem decide which computation method is the best.

Choose a computation method. Then solve.

At a book fair, 76 books were sold on the first day and 82 books were sold on the second day. How many books were sold during the two days?

Number of books sold on the first day + Number of books sold on the second day

        76   +   82

The numbers are small and 82 is close to a multiple of 10. You can use mental math.

$(76 + 2) + (82 - 2) = 78 + 80 = 158$

During the two days, 158 books were sold.

Choose a computation method. Then solve.

1. Of the 248 books on display, 46 were nonfiction books. How many books were not nonfiction books?

<u>mental math, 202 books</u>

2. Lisa bought 2 biographies for $5.37 each, a novel for $7.95, and a bookmark for $1.19. How much money did Lisa spend?

<u>paper and pencil, $19.88</u>

3. Over two days, 234 students visited the book fair in groups of 18. How many groups visited the fair?

<u>calculator, 13 groups</u>

48   **Holt Mathematics**

Challenge
1-6 *Finger Math*

Chisenbop is an ancient method of computation using your fingers. One of the best-known forms of Chisenbop is used for basic multiplication computations. It works only when all the factors are greater than 5. Follow these steps to use this form of the Chisenbop method of computation. The product of 6 × 7 is shown as an example.

Step 1 Subtract 5 from the first factor. Turn down that number of fingers on your left hand.

Step 2 Subtract 5 from the second factor. Turn down that number of fingers on your right hand.

Step 3 Multiply the total number of turned-down fingers on both hands by 10.

Step 4 Find the product of the numbers of fingers that are **not** turned down on each hand.

Step 5 Add the two products from Step 3 and Step 4.

$3 × 10 = 30$
↑
both hands

$4 × 3 = 12$
↑   ↑
left hand   **right hand**

$30 + 12 = 42$

So, $6 × 7 = 42.$

Use the Chisenbop method to find each product.

1. $7 × 8 = $ <u>56</u>   **2.** $6 × 9 = $ <u>54</u>   **3.** $8 × 6 = $ <u>48</u>

4. $8 × 9 = $ <u>72</u>   **5.** $6 × 6 = $ <u>36</u>   **6.** $9 × 7 = $ <u>63</u>

7. $7 × 7 = $ <u>49</u>   **8.** $7 × 9 = $ <u>63</u>   **9.** $8 × 8 = $ <u>64</u>

10. $6 × 8 = $ <u>48</u>   **11.** $9 × 9 = $ <u>81</u>   **12.** $9 × 8 = $ <u>72</u>

13. When would you choose to use the Chisenbop method of computation? When would you choose not to use that method? Explain.

<u>Possible answer: I would use this method when both factors are greater</u>
<u>than 5 and less than 11. I would not use this method when at least one of</u>
<u>the factors is less than or equal to 5.</u>

49   **Holt Mathematics**

Problem Solving
1-6 *Choose the Method of Computation*

Use the table below to answer questions 1–6. For each question, write the method of computation you should use to solve it. Then write the solution.

1. How many bones are in an average person's arms and hands altogether?

<u>mental math; 60 bones</u>

2. How many more bones are in an average person's head than chest?

<u>mental math; 3 bones</u>

3. Which part of the body has twice as many bones as the spine?

<u>mental math; feet</u>

4. How many bones are in the body altogether?

<u>paper and pencil; 206 bones</u>

5. A newborn baby has 350 bones. How many more bones does a newborn baby have than an adult?

<u>paper and pencil; 144 bones</u>

Bones in the Human Body

Body Part	Number of Bones
Head	28
Throat	1
Spine	26
Chest	25
Shoulders	4
Arms	6
Hands	54
Legs	10
Feet	52

6. How many bones are in each of an average person's feet, hands, legs, and arms?

<u>paper and pencil; feet: 26 bones;</u>
<u>hands: 27 bones; legs: 5 bones;</u>
<u>arms: 3 bones</u>

Choose the letter for the best answer.

7. The body's longest bones— thighbones and shinbones—are in the legs. The average thighbone is about 20 inches long, and the average shinbone is about 17 inches long. What is the total length of those four bones?

  Ⓐ paper and pencil; 74 inches
  B paper and pencil; 37 inches
  C mental math; 20 inches
  D calculator; 17 inches

8. The body has 650 muscles. Seventeen of those muscles are used to smile and 42 muscles are used to frown. How many more muscles are used to frown than to smile?

  F mental math; 35 muscles
  Ⓖ mental math; 25 muscles
  H paper and pencil; 608 muscles
  J calculator; 633 muscles

50   **Holt Mathematics**

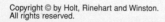

Holt Mathematics

Reading Strategies
Analyze Information

Read each problem. Then use the four steps to help you solve each problem.

> Marta kept track of the points she earned on 5 math tests. They were 85, 76, 88, 78, and 91. The total number of points possible on all 5 tests is 500 points. How many points has Marta earned so far?

1. What question is asked?

How many points has Marta earned?

2. What information is needed to answer the question?

Marta's test scores: 85, 76, 88, 78, and 91

3. Circle the operation needed to solve the problem.

(• Addition)   • Subtraction   • Multiplication   • Division

4. Show how you compute and solve the problem.

85 + 76 + 88 + 78 + 91 = 418 points

> Marta spends 35 minutes each night on her math homework. She spends another 45 minutes each night on the rest of her homework. How much time does Marta spend studying math over 5 days?

5. What question is asked?

How much time does Marta spend studying math in 5 days?

6. What information is needed to answer the question?

35 minutes each night on math; 5 nights of studying

7. Circle the operation needed to solve the problem.

• Addition   • Subtraction   (• Multiplication)   • Division

8. Show how you compute and solve the problem.

5 × 35 = 175 minutes

51   **Holt Mathematics**

Puzzles, Twisters & Teasers
Crossword Mania

Complete the crossword puzzle.

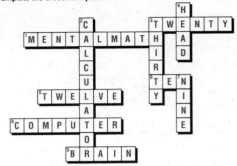

Across

1. Good computation method to use to solve number 3 down.

3. The number of seconds in a week is how much greater than 604,780?

5. The number of months in a century is how many times greater than the number of years in a century?

6. 121,780 is how many times greater than 12,178?

8. You can use this machine and the software that comes with it to do complex calculations. It can even help people do their taxes.

9. The best calculating tool humans have; it helps you do mental math.

Down

2. Good computation method to solve number 3 across.

3. A cheetah can run 60 miles an hour. If it could maintain that speed for half an hour, how many miles could it run?

4. You can find the answer to number 5 across inside this.

7. 56 × 42 is how much less than 2,361?

52   **Holt Mathematics**

Practice A
Patterns and Sequences

Choose the sequence that matches each pattern.

1. Start with 12; subtract 2.

A 2, 4, 6, 8, 10, 12, …
B 12, 11, 10, 9, 8, 7, …
C 12, 14, 16, 18, 20, …
(D) 12, 10, 8, 6, 4, 2, …

2. Start with 3; multiply by 2.

F 3, 5, 7, 9, 11, …
(G) 3, 6, 12, 24, 48, …
H 2, 6, 18, 54, 162, …
J 3, 4, 5, 6, 7, 8, 9, …

3. Start with 5; add 4.

A 5, 4, 3, 2, 1, …
(B) 5, 9, 13, 17, 21, …
C 5, 10, 15, 20, 25 …
D 4, 9, 14, 19, 24, …

4. Start with 1; multiply by 10.

(F) 1, 10, 100, 1,000, …
G 1, 10, 20, 30, 40, …
H 10, 100, 1,000, 10,000, …
J 10, 20, 30, 40, 50, …

Identify a pattern in each sequence.

5. 1, 4, 7, 10, 13, …

add 3

6. 15, 13, 11, 9, …

subtract 2

7. 5, 10, 15, 20, …

add 5

8.

Position	1	2	3	4	5	6	7
Value of Term	10	20	30	40	50	60	70

add 10

Name the next three terms in each sequence.

9. 1, 6, 11, 16, 21, ☐, ☐, ☐, …

26, 31, 36

10. 2, 4, 6, 8, 10, ☐, ☐, ☐, …

12, 14, 16

11.

Position	1	2	3	4	5	6	7
Value of Term	50	45	40	35	30	25	20

15; 10; 5

12. The temperature was 79°F on Friday, 76°F on Saturday, and 73°F on Sunday. If this weather pattern continues, what will the temperature be on Monday?

70°F

13. Tony's Cafe sells four sizes of pizza. The first three sizes are 8 inches, 10 inches, and 12 inches. If this pattern of size continues, what is the largest size pizza the cafe sells?

14 inches

53   **Holt Mathematics**

Practice B
Patterns and Sequences

Identify a pattern in each sequence and then find the missing terms.

1. 4, 8, 16, 32, ☐, ☐, ☐, …

multiply by 2; 64, 128, 256

2. 100, 95, 90, 85, ☐, ☐, ☐, …

subtract 5; 80; 75; 70

3. 8, 20, 32, 44, ☐, ☐, ☐, …

add 12; 56; 68; 80

4. 6, 12, 18, 24, ☐, ☐, ☐, …

add 6; 30; 36; 42

5. 9, 18, 27, 36, ☐, ☐, ☐, …

add 9; 45, 54, 63

6. 3, 6, 12, 24, ☐, ☐, ☐, …

multiply by 2; 48, 96, 192

7.

Position	1	2	3	4	5	6	7
Value of Term	5	10	20	40	80	160	320

multiply by 2

8. 300, 250, ☐, ☐, 100, ☐, 0, …

subtract 50; 200; 150; 50

9. 1, 15, ☐, 43, 57, ☐, 85, 99, …

add 14; 29; 71

10. 7, ☐, 21, 28, ☐, ☐, ☐, 56, …

add 7; 14; 35; 42; 49

11. 9, ☐, 13, ☐, ☐, ☐, 21, 23, …

add 2; 11; 15; 17; 19

12.

Position	1	2	3	4	5	6	7
Value of Term	3	12	21	30	39	48	57

add 9

13. A forest ranger in Australia took measurements of a eucalyptus tree for the past 3 weeks. The tree was 12 inches tall the first week, 19 inches the second week, and 26 inches the third week. If this growth pattern continues, how tall will the tree be next week?

33 inches

14. Maria puts the same amount of money in her savings account each month. She had $450 in the account in April, $600 in May, and $750 in June. If she continues her savings pattern, how much money will she have in the account in July?

$900

54   **Holt Mathematics**

Identify a pattern in each sequence, and then find the missing terms.

1. 13, 17, 21, 25, □, □, □, ...

add 4; 29, 33, 37

2. 48, 45, 42, 39, □, □, □, ...

subtract 3; 36; 33, 30

3. 1,600, 800, 400, 200, □, □, □, ...

divide by 2; 100, 50, 25

4. 1, 3, 9, 27, □, □, □, ...

multiply by 3; 81; 243; 729

5.

Position	1	2	3	4	5	6	7
Value of Term	19	38	57	76	95	114	133

add 19

Identify a pattern in each sequence. Name the missing terms.

6. 2, □, □, 16, 32, □, ...

multiply by 2; 4, 8, 64

7. 8, □, 24, □, 40, □, 56, ...

add 8; 16; 32; 48

8.

Position	1	2	3	4	5	6	7
Value of Term	4	12	14	42	44	132	134

multiply by 3 and then add 2

9. The sequence 1, 1, 2, 3, 5, 8, 13, ... is called the Fibonacci Sequence. Identify the pattern in the sequence, and name the next three terms.

The next number in the sequence is the sum of the two previous

numbers; 21, 34, 55

10. The pattern below shows a special sequence called triangular numbers. Draw the next figure in the pattern, and name the next triangular number in the sequence.

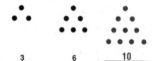

1   3   6   _10_

Find the next three numbers in the sequence.

8, 12, 16, 20, 24, □, □, □, ...

Step 1: Look at pairs of numbers to find the pattern.

8, 12, 16, 20, 24, □, □, □, ...

8 + 4 = 12   12 + 4 = 16   16 + 4 = 20

The pattern is to add 4.

Step 2: Use the pattern to name the next three numbers.

24 + 4 = 28   28 + 4 = 32   32 + 4 = 36

The next three numbers are 28, 32, and 36.

Find the next three numbers in each sequence.

1. 5, 8, 6, 9, 7 □, □, □, ...

_10, 8, 11_

2. 90, 80, 70, 60, □, □, □, ...

_50, 40, 30_

3. 2, 4, 8, 16, □, □, □, ...

_8, 32, 16_

4. 10, 14, 18, 22, □, □, □, ...

_26, 30, 34_

5. 13, 21, 29, 37, □, □, □, ...

_45, 53, 61_

6. 24, 12, 16, 8, □, □, □, ...

_12, 6, 10_

7. 14, 12, 10, 8, □, □, □, ...

_6, 4, 2_

8. 1, 7, 13, 19, □, □, □, ...

_25, 31, 37_

9. 1, 3, 6, 10, □, □, □, ...

_15, 21, 28_

10. 40, 38, 36, □, □, □, ...

_34, 32, 30_

11. 54, 45, 36, □, □, □, ...

_27, 18, 9_

12. 10, 25, 40, □, □, □, ...

_55, 70, 85_

13. 36, 29, 22, □, □, □, ...

_15, 8, 1_

14. 18, 36, 72, □, □, □, ...

_144, 288, 576_

Draw the next three figures in each pattern.

1.

2.

3.

4.

5.

6.

7.

---

1. A giant bamboo plant was 5 inches tall on Monday, 23 inches tall on Tuesday, 41 inches tall on Wednesday, and 59 inches tall on Thursday. Describe the pattern. If the pattern continues, how tall will the giant bamboo plant be on Friday, Saturday, and Sunday?

Each day the giant bamboo plant grew 18 inches. The giant

bamboo plant will be 77 inches tall on Friday, 95 inches tall

on Saturday, and 113 inches tall on Sunday.

2. A scientist was studying a cell. After the second hour there were two cells. After the third hour there were four cells. After the fourth hour there were eight cells. Describe the pattern. If the pattern continues, how many cells will there be after the fifth, sixth, and seventh hour?

The number of cells doubled every hour. After the fifth hour

there will be 16 cells. After the sixth hour there will be 32 cells.

After the seventh hour there will be 64 cells.

Choose the letter for the best answer.

3. The first place prize for a sweepstakes is $8,000. The third place prize is $2,000. The fourth place prize is $1,000. The fifth place prize is $500. What is the second place prize?

A $7,000   Ⓒ $4,000

B $6,000   **D** $3,000

4. The temperature was 59°F at 3:00 A.M., 62°F at 5:00 A.M., and 65°F at 7:00 A.M. If the pattern continues, what will the temperature be at 9:00 A.M., 11:00 A.M., and 1:00 P.M.?

F 66°F at 9:00 A.M., 67°F at 11:00 A.M., 68°F at 1:00 P.M.

G 68°F at 9:00 A.M., 70°F at 11:00 A.M., 72°F at 1:00 P.M.

Ⓗ 68°F at 9:00 A.M., 71°F at 11:00 A.M., 74°F at 1:00 P.M.

J 70°F at 9:00 A.M., 75°F at 11:00 A.M., 80°F at 1:00 P.M.

Reading Strategies
Sequence

When a story has several parts, such as *Star Wars*, you want to watch them in order, or in **sequence.** When a set of numbers is **in order,** it is called a **sequence.** A sequence of numbers may have a special pattern. When we talk about the pattern, each number in the sequence is called a **term.**

Read each term of the sequence below. Note how the numbers change from one term to the next. What do you think is the pattern in this sequence?

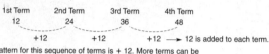

1st Term   2nd Term   3rd Term   4th Term
  12   24   36   48
     +12   +12   +12 → 12 is added to each term.

The pattern for this sequence of terms is + 12. More terms can be added to a sequence by continuing the pattern.

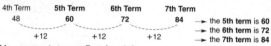

4th Term   5th Term   6th Term   7th Term
  48   60   72   84 → the **5th term is 60**
     +12   +12   +12 → the **6th term is 72**
                                       → the **7th term is 84**

Use this sequence to answer Exercises 1–6:
  78 77 75 72 68

1. What is the 3rd term in the sequence? _____ 75

2. What change occurs between the 1st term and 2nd term?

The 2nd term is one less than the 1st term.

3. What change occurs between the 2nd term and the 3rd term?

The 3rd term is 2 less than the 2nd term.

4. What change occurs between the 3rd term and the 4th term?

The 4th term is 3 less than the 3rd term.

5. Write a rule to describe how to find each term in the sequence.

Subtract 1 more than you subtracted from the previous

term: −1, −2, −3, and so on.

6. Identify the 6th, 7th, and 8th terms in this sequence.

63, 57, 50

59   **Holt Mathematics**

Puzzles, Twisters & Teasers
Have You Heard?

Did you hear about the student who tried using a broken calculator to evaluate $29^7 + 13^4 + 5^8$?

Complete each sequence. Use the decoder below to find the letter that corresponds to the answer for each exercise. Place each letter above it's corresponding exercise number below to complete the answer to the puzzle.

1. 3, 6, 9, 12, __15__

2. 8, 10, 12, 14, 16, __18__

3. 8, 16, __24__, 32, 40, 48

4. __5__, 10, 15, 20, 25, 30

5. 20, 16, 12, 8, 4, __0__

6. 0, 6, __12__, 18, 24, 30

7. __22__, 20, 18, 16, 14, 12

8. 1, 2, 4, 8, 16, __32__

9. 7, 14, 21, __28__, 35

10. 3, __6__, 12, 24, 48

0 = B	18 = S
5 = Y	22 = E
6 = C	24 = A
12 = O	28 = I
15 = N	32 = D

He couldn't even reach   S   E   C   O   N   D
                              2   7   10   6   1   8

                              B   A   S   E .
                              5   3   2   7

60   **Holt Mathematics**
